THE DEEP MIXING METHOD
Principle, Design and Construction

The Deep Mixing Method
Principle, Design and Construction

Edited by
COASTAL DEVELOPMENT INSTITUTE OF
TECHNOLOGY (CDIT), JAPAN

A.A. BALKEMA PUBLISHERS / LISSE / ABINGDON / EXTON (PA) / TOKYO

Library of Congress Cataloging-in-Publication Data

Applied for

Cover design: Studio Jan de Boer, Amsterdam, The Netherlands.
Printed by: Grafisch Produktiebedrijf Gorter, Steenwijk, The Netherlands.

Copyright © 2002 Swets & Zeitlinger B.V., Lisse, The Netherlands

All rights reserved. No part of this publication or the information contained herein may be reproduced, stored in a retrieval system, or transmitted in any form or by any means, electronic, mechanical, by photocopying, recording or otherwise, without written prior permission from the publishers.

Although all care is taken to ensure the integrity and quality of this publication and the information herein, no responsibility is assumed by the publishers nor the author for any damage to property or persons as a result of operation or use of this publication and/or the information contained herein.

Published by: A.A. Balkema Publishers, a member of Swets & Zeitlinger Publishers
              www.balkema.nl and www.szp.swets.nl

ISBN 90 5809 367 0

Contents

Preface vii

List of Committee Members ix

List of Technical Terms and Symbols xi

Chapter 1 Outline of the Deep Mixing Method 1
   1.1 Introduction 1
   1.2 The development of the Deep Mixing Method 4
   1.3 Types of stabilizing agents and basic mechanisms of stabilization .. 8
   1.4 Scope of the text 11

Chapter 2 Factors Affecting Strength Increase 17
   2.1 General 17
   2.2 Influence of various factors on the strength of lime treated soil .. 18
   2.3 Influence of various factors on the strength of cement treated soil .. 24
   2.4 Prediction of strength 33

Chapter 3 Engineering Properties of Treated Soils 37
   3.1 Introduction 37
   3.2 Physical properties 37
   3.3 Mechanical properties (strength characteristics) 40
   3.4 Mechanical properties (consolidation characteristics) ... 45
   3.5 Engineering properties of cement treated soil manufactured in-situ .. 49

Chapter 4 Applications 53
   4.1 Patterns of applications 53
   4.2 Applications in Japan 59

Chapter 5 Design of Improved Ground by DMM 69
  5.1 Introduction 69
  5.2 Design procedure for block-type and wall-type improvements 70
  5.3 Design procedure for group column-type improvement 82

Chapter 6 Construction Procedures and Control 93
  6.1 Classification of techniques 93
  6.2 Marine works 93
  6.3 On land works 101
  6.4 Quality control and quality assurance 107

Appendix A Summary of the Practice for Making and Curing Stabilized
            Soil Specimens without Compaction (JGS 0821) 111

Appendix B Influence of In-situ Mixing Conditions on the Quality
            of Treated Soil 113

Appendix C Recent Research Activity on the Group Column-Type
            Improved Ground 119

Preface

Due to the growing population and spreading urbanization in the past century, it has become really difficult to select the suitable ground for locating infrastructures. Construction projects often encounter very soft soil deposits, which will pose headache problems of stability and/or excessive settlement. To solve these problems, a variety of ground improvement techniques have been developed and put into practice so far. The Deep Mixing Method (DMM), a deep in-situ soil stabilization technique using cement and/or lime as a stabilizing agent, was developed in Japan and in the Scandinavian countries independently in 1970s. Numerous research efforts have been paid in these areas to investigate properties of treated soil, behavior of DMM improved ground under static and dynamic conditions, design method, and execution techniques. Today, in each year, roughly three to four million cubic meters of improvement is done by DMM in Japan alone and around two million meters of lime columns are installed in Scandinavian Countries. To enhance the exchange of information between Japan, Scandinavian countries and the other parts of the world, the Japanese Geotechnical Society and ISSMGE TC-17 jointly initiated the specialty international conference on Deep Mixing in 1996 in Tokyo. It was followed by two international symposia in 1999 in Stockholm and 2000 in Helsinki and also by a couple of workshops in the USA.

This book is a concise English version of a manual on DMM written in Japanese and published in 1999 by the Coastal Development Institute of Technology, Japan (CDIT), an organization under the jurisdiction of the Japanese Ministry of Land, Infrastructure and Transport. A list of the committee members for editing the English version is indicated on a consequent page. Translation and editing of the English version is financed by CDIT.

The book presents properties of the treated soil, various applications, design and execution of DMM, that are based on a lot of research efforts and accumulated experience in Japan. As the soils are local materials and the applications are different in various parts of the world, the book is not aimed to provide a complete design manual of DMM but is rather a state of practice of the technology. In the actual design and construction, it is always advisable to employ the updated engineering expertise and know-hows.

We hope that this book will be a useful reference for engineers and researchers involved in the Deep Mixing Method around the world.

April 2001

Masaki Kitazume
Masaaki Terashi

List of Committee Members

Kitazume, Masaki	Port and Airport Research Institute
Katsuumi, Tsutomu	Coastal Development Institute of Technology
Mori, Hirofumi	Coastal Development Institute of Technology
Nakamura, Toshitomo	Cement Deep Mixing Method Association
Nozu, Mitsuo	Cement Deep Mixing Method Association
Oono, Yasutoshi	Cement Deep Mixing Method Association
Terashi, Masaaki	Nikken Sekkei Nakase Geotechnical Institute
Tokunaga, Sachihiko	Cement Deep Mixing Method Association
Yamamoto, Yoshio	Cement Deep Mixing Method Association
Yokoyama, Katsuhiko	Cement Deep Mixing Method Association

List of Technical Terms and Symbols

additive	chemical material to be added to stabilizing agent for improving characteristics of treated soil
cement slurry	slurry-like mixture of cement and water
DM machine	a machine to be used to construct DM column
external stability	overall stability of the stabilized body
fixed type	a type of improvement in which treated soil column reaches a bearing layer
floating type	a type of improvement in which treated soil column ends in a soft soil layer
improved ground	a region with stabilized body and surrounding original soil
internal stability	stability on internal failure of improved ground
original soil	soil left without the treatment
stabilizing agent	chemically reactive materials (lime, cement, etc)
stabilized body	a sort of underground structure constructed by the stabilized columns
stabilized column	a column of treated soil constructed by a single operation of deep mixing machine
surrounding ground	soil surrounding the improved ground
treated soil	soil treated by mixing with stabilizing agent
aw	amount of stabilizing agent, defined as a ratio of the dry weight of stabilizing agent to the dry weight of original soil.
B	width of improved ground (m)

D	depth of improved ground, or vertical length of long wall in wall type improved ground (m)
Fs	factor of safety
L_l	width of long wall in wall type improved ground (m)
L_S	width of short wall in wall type improved ground (m)
qu	unconfined compressive strength (kN/m²)
qu_{ck}	design unconfined compressive strength (kN/m²)
qu_l	unconfined compressive strength of laboratory treated soil (kN/m²)
qu_f	unconfined compressive strength of in-situ treated soil (kN/m²)
qu_n	unconfined compressive strength after n-days' curing (kN/m²)
R_l	a ratio of long wall breadth in wall type improved ground (m)
R_S	a ratio of short wall breadth in wall type improved ground (m)
w_n	natural water content of original soil
w_i	initial water content of untreated soil before stabilizing
α	coefficient of effective width of treated soil column
α	stabilizing factor, dry weight of stabilizing agent per 1 m³
β	reliability coefficient of over-lapping
γ	correction factor for scattered strength
λ	ratio of qu_f and qu_l
τ_a	allowable shear strength (kN/m²)
σ_{ca}	allowable compressive strength (kN/m²)
σ_t	allowable tensile strength (kN/m²)

Lineup of Deep Mixing barges with eight mixing shafts
- DECOM #7, DCM #3, POCM #2 (from left to right)
- Improved cross section is 5.7m^2

Deep Mixing machines for on-land works with tower height of over 40m
- Deep Mixing machines are ordinarily with two mixing shafts

Exposed sea bottom improved by Wet type Deep Mixing Method at the intake of Aioi thermoelectric power station

Retaining Wall by Deep Mixing columns
- Fly ash cement mixture is used as the stabilizer

Excavated surface of Retaining Wall by Deep Mixing columns

CHAPTER 1

Outline of the Deep Mixing Method

1.1 INTRODUCTION

It is an obvious truism that, structures should be constructed on good quality ground. The ground conditions of construction sites, however, have become worse than ever during recent decades throughout the world. This situation is especially pronounced in Japan, where many construction projects are conducted on soft alluvial clay ground, land reclaimed with dredged soils, highly organic soils and so on. Figures 1.1 and 1.2 show the typical physical characteristics of soft soils often encountered in marine and on land constructions respectively. In the figures, the relationship between plasticity index (I_p) and liquid limit (w_l) is plotted. Most clayey soils with a high liquid limit cause both stability and deformation problems during and after construction.

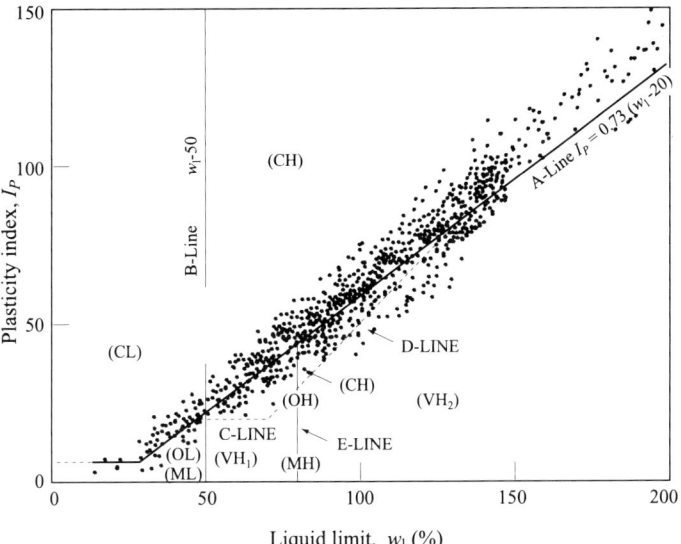

Figure 1.1. Plasticity of soft ground in Japanese coastal area (Ogawa & Matsumoto, 1978).

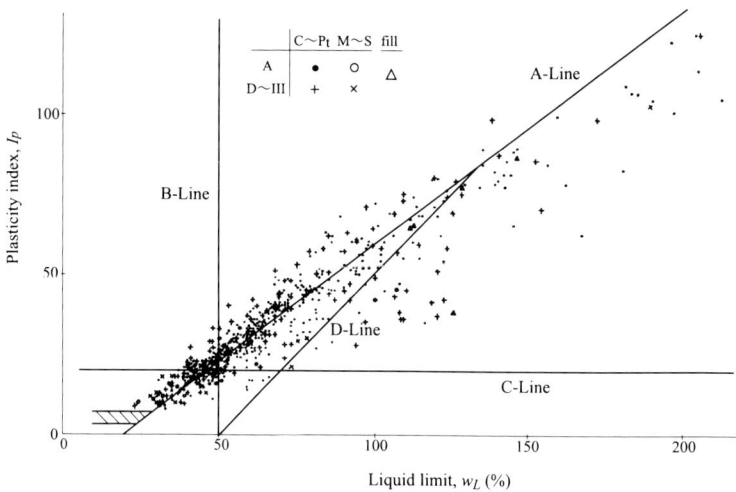

Figure 1.2. Plasticity of soft ground encountered in Japanese railway construction (Watanabe, 1974).

Apart from these clayey or highly organic soils, loose sand deposits under the water table cause serious problems of liquefaction under seismic conditions. In order to cope with these problems, various kinds of ground improvement methods have been introduced to or originally developed in Japan during the past decades as shown in Table 1.1 (Ministry of Transport, 1980). The Deep Mixing Method (DMM) that will be described in the current text is categorized as a solidification (or admixture stabilization) method.

DMM utilizes such stabilizing agents as quicklime, slaked lime, cement, or a combination of these agents. The concept of lime stabilization has a long history. The in-situ mixing approach may also have a variety of roots in the past. However, the research and development of current DMM began in the late 1960s using lime as a stabilizing agent. DMM was put into practice in Japan and Nordic countries in the middle of the 1970s, and then spread to China, South East Asia, and recently to other parts of the world, including the USA. More than two decades of practice have seen the equipment improved, stabilizing agents changed, and the applications diversified. Lime has been replaced with cement in Japan, and with a lime-cement mixture in Nordic countries. More recently, materials combining lime or cement with gypsum, fly ash and slag have appeared and are employed for particular purposes. Figure 1.3 shows the Mark III Deep Mixing machine in the early 1970s. Figure 1.4 is an example of a recent marine application of the Deep Mixing Method at the construction site of the Trans-Tokyo Bay Highway.

Table 1.1. Principal soft ground improvement methods in Japan.

Improvement Principle	Engineering Method	Work Examples	Period Practical Application Introduced
Replacement	Excavation Replacement	Dredging replacement method	1930s →
Replacement	Forced replacement	Sand compaction pile method	1966 →
Consolidation	Preloading	Preloading method	1928 →
Consolidation	Preloading with vertical drain	Sand drain method	1952 →
Consolidation	Preloading with vertical drain	Packed sand drain method	1967 →
Consolidation	Preloading with vertical drain	Board drain method	1963 →
Consolidation	Dewatering	Deep well method	1944 →
Consolidation	Dewatering	Well point method	1953 →
Consolidation	Dewatering	Vacuum consolidation method	1971 →
Consolidation	Chemical dewatering	Quick lime pile method	1963 →
Increasing density / Dewatering / compaction	Compaction by displacement and vibration	Sand compaction pile method	1957 →
Increasing density / Dewatering / compaction	Compaction by displacement and vibration	Gravel compaction pile method	1965 →
Increasing density / Compaction	Vibration compaction	Vibro-flotation method	1955 →
Increasing density / Compaction	Impact compaction	Dynamic consolidation method	1973 →
Solidification (Admixture stabilization)	Agitation mixing	Shallow mixing method	1972 →
Solidification (Admixture stabilization)	Agitation mixing	Deep mixing method	1974 →
Solidification (Admixture stabilization)	Jet mixing	Jet mixing method	1981 →
Contact pressure reduction	Load distribution	Fascine mattress method	1930s →
Contact pressure reduction	Load distribution	Sheet net method	1962 →
Contact pressure reduction	Load distribution	Sand mat method	1930s →
Contact pressure reduction	Load distribution	Surface solidification method	1970 →
Contact pressure reduction	Balancing loads	Counterweight fill method	1930s →

4 *The Deep Mixing Method – Principle, Design and Construction*

Figure 1.3. First full-scale 1970's test at offshore Haneda.

Figure 1.4. Marine work using Deep Mixing Method on Trans-Tokyo Bay Highway.

1.2 THE DEVELOPMENT OF THE DEEP MIXING METHOD

(1) *historical review of R&D in Japan*. The development of the Deep Mixing Method in Japan was briefly reviewed by Terashi in his theme lecture at the ISSMFE Hamburg Conference (Terashi, 1997). Research and development of the Deep Mixing Method in Japan was initiated by the Soil Stabilization Laboratory at the Port and Harbour Research Institute (PHRI) of the Japanese Ministry of Transport. The concept of lime stabilization of marine clays was first publicized in a technical publication of the PHRI in 1968 (Yanase, 1968). The concept was realized by the research and development studies of Okumura, Terashi and their colleagues in the early 1970s. The early research was conducted to achieve two

goals: one was to investigate the lime reactivity of a variety of Japanese marine clays and the other was to develop equipment which permits a constant supply of stabilizing agent and reasonably uniform mixing at depth. Most of the marine clays tested easily gained strength of the order of 100 kN/m² to 1 MN/m² in terms of unconfined compressive strength.

The development of the equipment (Mark I to Mark III) was done at the PHRI with the collaboration of Toho Chika Koki Co. Ltd. The first field trial test was done with the Mark II machine, which was only 2 m high. The first trial on the sea was done near-shore at Haneda Airport with the Mark III, which was capable of improving the sea bottom sediment up to 10 meters from the water surface. The basic mechanism of the equipment was established by these trials. Finally the Mark IV machine was manufactured by Kobe Steel Co. Ltd. and a marine trial test was done by the PHRI near-shore at Nishinomiya to establish the construction control procedure. These steps in the initial development of the method were continuously publicized through annual meetings of the Japanese Geotechnical Society and later through the PHRI reports (e.g. Okumura et al., 1972). Okumura and Terashi introduced the technology to the international community in 1975 at the ISSMFE 5th Asian Regional Conference (Okumura & Terashi, 1975). Stimulated by these activities in the development of the new technique, a number of Japanese contractors started their own research and development of this technique in the middle of the 1970s.

As granular quicklime or powdered slaked lime was used as a stabilizing agent in these initial development stages, the method was named "Deep Lime Mixing (DLM)". The first contractor who put DLM into practice was Fudo Construction Co. Ltd. The very first application was the use of the Mark IV machine to improve reclaimed soft alluvial clay in Chiba prefecture in June 1974. In the five years before 1978, DLM was practiced at twenty-one construction sites including two marine works. In an effort to improve the uniformity of the improved soil, cement mortar and cement slurry quickly replaced granular quicklime. A method using a variety of stabilizing agents in slurry form is now called the Cement Deep Mixing (CDM) method and, more generally, is referred to as the Wet Method of Deep Mixing Method.

During the 1970s and 80s, Terashi and Tanaka at the PHRI continued to study the engineering properties of lime and cement improved soils (Terashi et al., 1979, Terashi et al., 1983c) and proposed a laboratory mold test procedure. Essentially the same procedure was later standardized by the Japanese Geotechnical Society in 1990 and experienced a minor revision in 2000 (Japanese Geotechnical Society, 2000). Terashi, Tanaka and Kitazume extended the study to investigate the behavior of improved ground (Terashi & Tanaka, 1981, 1983, Terashi et al., 1983b). During this period in the early 1980s, the Japanese Geotechnical Society (JGS) established a technical committee to compile the State of the Art DMM and the essence was reported in the monthly journal of the Society (Terashi et al., 1983a, Noto et al., 1983, Terashi, 1983). In 1983 the Ministry of Transport established a working group comprising engineers from local port construction bureaus and the PHRI, which spent three years compiling the full details of the design procedure and case histories (Ministry of Transport, 1986). The PHRI also conducted a joint research program on the same lines with Takenaka Doboku. These efforts were summarized in the CDM technical manual (Cement Deep Mixing Method Association, 1993). The JGS and ISSMFE TC-17 co-sponsored an international conference: IS-Tokyo '96. The proceedings of the conference, Grouting and Deep Mixing, contain detailed information of the practices of this method (Yonekura et al., 1996). The latest technical manual for the wet method of deep mixing was published in 1999 (Coastal Development Institute of Technology, 1999).

A research group at the Public Works Research Institute of the Japanese Ministry of Construction started studies to develop a similar technique from the late 1970s to the early 1980s, inviting staff of the PHRI to take part as advisory committee members. The technique developed is called the Dry Jet Mixing (DJM) Method in which dry powdered cement or lime is used as a stabilizing agent instead of slurry. This is the so-called "Dry Method of DMM". An outline of the method as well as design and execution procedures is described in a recent technical manual (Public Works Research Center, 1999).

Since a variety of equipment was established and a standard design procedure became available, the application has exploded. Figure 1.5 shows statistics in Japan. As shown in the figure, the wet method (CDM) is preferred in marine work while both the dry (DJM) and wet (CDM) methods are employed for on land work. The total volume of treatment up to 1999 amounts to 61 million m^3 in Japan.

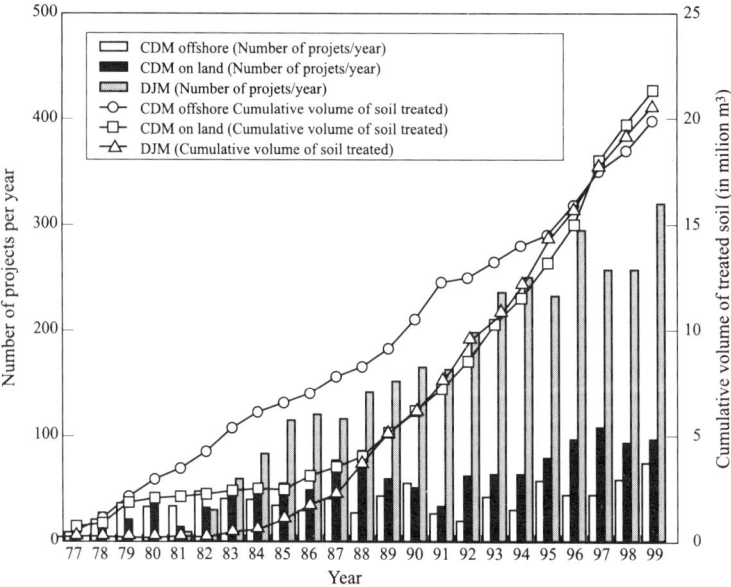

Figure 1.5. Cumulative volume of soil treated by deep mixing and number of projects on annual basis in Japan.

(2) *historical review of R&D in Nordic countries*. The development of the Deep Mixing Method in Nordic countries is summarized by Rathmayer in his State of the Art Report (Rathmayer, 1996) to IS-Tokyo '96. In 1967 a new method for stabilizing soft clay by quicklime was developed on the initiative of Kjeld Paus. The method was named the Swedish Lime Column method. Linden-Alimak AB developed light wheel-mounted mixing equipment in co-operation with the Swedish Geotechnical Institute, the construction enterprise BPA Byggproduktion AB and the company Euroc AB. A careful description of the Swedish method was first prepared by Assarson et al. in 1974 (Assarson et al., 1974). Broms and Boman (1975) reported this new technique to the international geotechnical community in 1975 at the 5th Asian Regional Conference on Soil Mechanics and Foundation Engineering, which was the same conference where Okumura and Terashi reported the Japanese equivalent. The first design handbook, written by Broms and Boman, was issued in 1977 (Broms & Boman, 1977).

In Nordic countries, the major purpose of improvement by lime columns is to reduce settlement at road embankments, bridge abutments, dwelling foundations and so on. In addition, the technique is applied to increase the stability of embankments, excavated slopes and so on. A mixture of lime and cement is also nowadays commonly used as a stabilizing agent in order to obtain relatively high strength. Figure 1.6 shows recent statistics for both Finland and Sweden (Rathmayer, 1996). In 1992, for example, the total linear meters of installation exceeded 1 million meters in both Finland and Sweden.

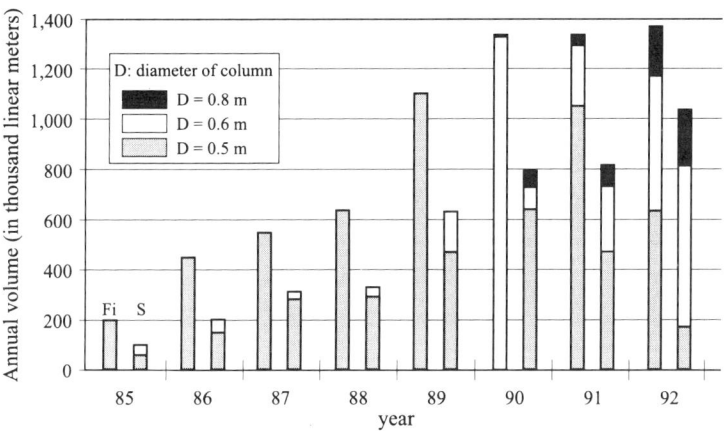

Figure 1.6. Annual volume (in linear meters) of deep mixing treatments in Finland (left bar) and Sweden (right bar).

1.3 TYPES OF STABILIZING AGENTS AND BASIC MECHANISMS OF STABILIZATION

The stabilizing agents used in practice are, in the majority of cases, Portland cement and lime, but dozens of stabilizing agents are now available on the Japanese market. Some of these newly developed special stabilizing agents are designed for the improvement of clay soils with high water content or organic soils, for which ordinary cement or lime is not very effective. Some other stabilizing agents are designed for cases where the rate of strength increase has to be controlled for the convenience of the construction. These stabilizing agents react slowly with soil and exhibit smaller strength in the short term, but result in sufficiently high strength in the long term in comparison with ordinary cement. The recent Recycling Act for the efficient utilization of industrial by-products has led to further development of stabilizing agents (for example, Asano et al., 1996). The basic mechanisms of admixture stabilization using cement or lime were extensively studied for several decades by highway engineers. This is because lime or cement treated soils have been used for sub-base or sub-grade materials in road constructions (Ingles & Metcalf, 1972). In the following paragraphs, the basic mechanisms of lime and cement stabilization are briefly reviewed (Babasaki et al., 1996).

(1) *lime-type stabilizing agents*. When mixed with soil, quicklime (CaO) absorbs moisture in the soil to become hydrated lime (CaO + H_2O = $Ca(OH)_2$). This hydration reaction is rapid and generates a large amount of heat. During the process, lime doubles in volume. The water content of soft soil is thus reduced, accompanied by a slight increase in shear strength. The process is a kind of consolidation successfully applied to the "quicklime pile method of soil improvement", which is a technique different from DMM.

With the existence of sufficient pore water, hydrated lime dissolves into water and increases the calcium and hydroxyl ion contents. Then Ca^{2+} ions exchange with cations on the surface of the clay minerals. The cation exchange reaction alters the characteristics of water films on the clay minerals. In general, the plastic limit (w_p) of the soil is increased, reducing the plasticity index. Furthermore, under a high concentration of hydroxyl ions (high pH condition), silica and/or aluminum in the clay minerals dissolve into the pore water and react with calcium to form a tough water-insoluble gel of calcium-silicate and/or calcium-aluminate. The reaction (so-called pozzolanic reaction) proceeds as long as the high pH condition is maintained and calcium exists in excess. The strength increase of lime treated soil is attributed mainly to the pozzolanic reaction product, which cements the clay particles together. The basic mechanism of lime stabilization is shown schematically by Ingles & Metcalf (1972) in Figure 1.7.

As is described above, the strength increase of lime treated soils relies solely upon a chemical reaction between clay and lime. The formation of cementation material commences after the attack of lime on clay minerals. Therefore, the

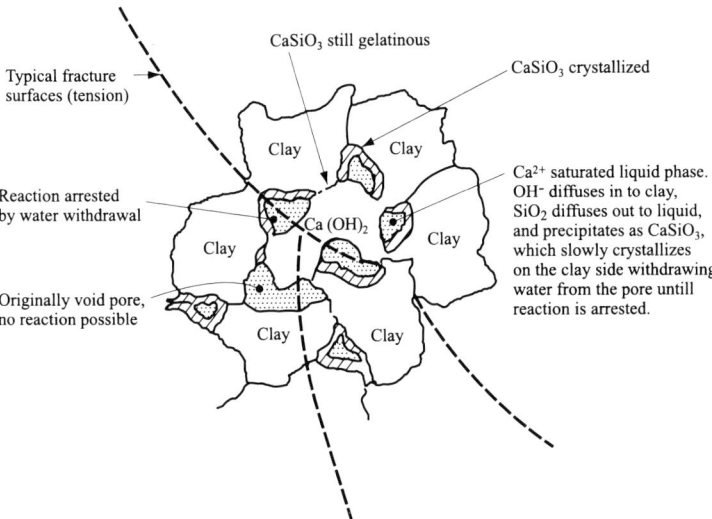

Figure 1.7. Basic lime stabilization mechanism (Ingles & Metcalf, 1972).

thorough mixing of soil and lime is absolutely necessary to increase the contact area of lime and clay.

(2) *cement-type stabilizing agents*.

(a) *standard cement-type stabilizing agents*. The cement types used as stabilizing agents are "Portland cement" or "Portland blast furnace slag cement". Portland cement is made by adding gypsum to cement clinker and grinding it to powder. Cement clinker is formed by minerals; $3CaO.SiO_2$, $2CaO.SiO_3$, $3CaO.Al_2O_3$ and $4CaO.Al_2O_3.Fe_2O_3$. A cement mineral, $3CaO.SiO_2$, for example, reacts with water in the following way to produce cement hydration products.

$$2(3CaO.SiO_2) + 6H_2O = 3CaO.2SiO_2.3H_2O + 3Ca(OH)_2 \cdots \text{Eq. (1.1)}$$

During the hydration of cement, calcium hydroxide is released. The cement hydration product has high strength, which increases as it ages, while calcium hydroxide contributes to the pozzolanic reaction as in the case of lime stabilization. Portland blast furnace slag cement is a mixture of Portland cement and blast furnace slag. Finely powdered blast furnace slag does not react with water but has the potential to produce pozzolanic reaction products under high alkaline conditions. In Portland blast furnace slag cement, the SiO_2 and Al_2O_3 contained in the slag are actively released by the stimulus of the large quantities of Ca^{2+} and SO_4^{2-} released from the cement, so that a fine hydration product abounding in silicates is formed rather than a cement hydration product, and the long-term strength is enhanced.

The rather complicated mechanism of cement stabilization is simplified and

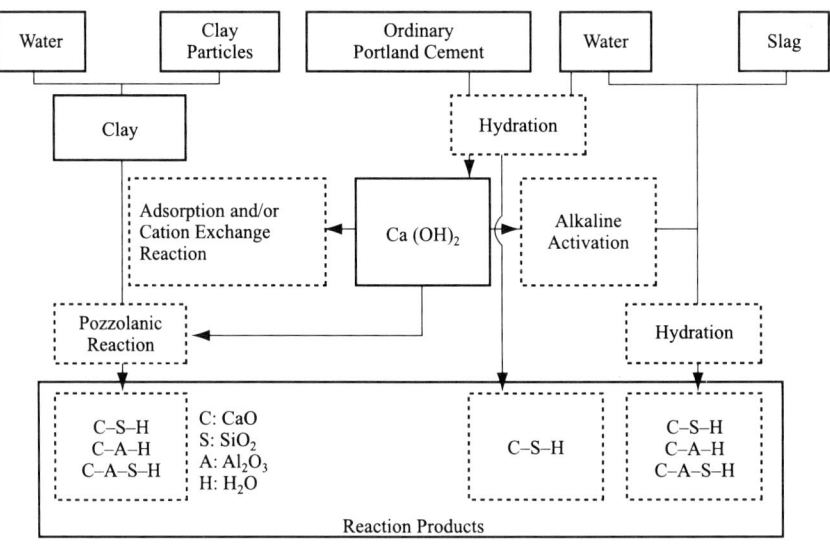

Figure 1.8. Chemical reactions between clay, cement, slag and water (Saitoh et al., 1985).

schematically shown in Figure 1.8 for the chemical reactions between clay, pore water, cement and slag (Saitoh et al., 1985).

(b) *special cement-type stabilizing agents*. Special cement-type stabilizing agents are cements that are specially prepared for the purpose of stabilizing soil or similar material by reinforcing certain constituents of the ordinary cement, by adjusting the grain size or by adding ingredients effective for particular soil types (Japan Cement Association, 1994). The improvement effect in organic soils is said to be affected by the ratio ((SiO_2 + Al_2O_3)/CaO) of the constituent elements in cement-type stabilizing agents (Hayashi et al., 1989). "Delayed stabilizing" or "long-term strength control" type stabilizing agents by which the rate of strength increase can be controlled, are obtained by adjusting the quantities of ingredients such as gypsum or lime.

(3) *difference in the strength increase with time between lime and cement*. Although the improvement by lime and cement is based on similar chemical reactions, the rate of strength increase differs. Figure 1.9 schematically compares the occurrence of chemical reactions with time in lime-soil mixture and cement soil mixture. In both cases, reduction of water content due to hydration precedes all other reactions if a dry-powder stabilizer is added. The reduction of water content leads to a slight increase of strength. Following a reaction common to both stabilizing agent is a cation exchange, which leads to an improvement in the plasticity of soils.

After these reactions, substantial hardening of the mixture starts. In the case of lime treatment, the pozzolanic reaction between lime and clayey soils is slow but

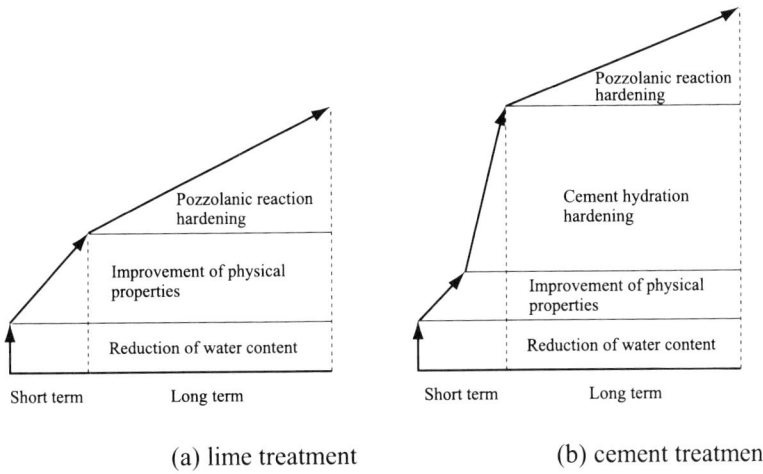

Figure 1.9. Simplified explanation of different time process of the improvement by lime and cement.

lasts for years. In contrast to this, in the case of cement treatment, the formation of cement hydration product is relatively rapid and most of the strength increase due to cement hydration is completed within several weeks. The lime liberated during the hydration of cement also reacts with claycy soils as well and the strength increase due to this also lasts for a long time in the case of cement treatment.

1.4 SCOPE OF THE TEXT

When we look at the strength of treated soil, CDM (the wet method) creates soil with strength exceeding 1 MN/m^2 in terms of unconfined compressive strength. DJM (the Japanese dry method) mostly employed in a group column type creates soil with 500kN/m^2. The Swedish lime columns are ordinarily used at a strength less than 150kN/m^2. The difference in the strength naturally causes differences in the relative stiffness of treated and untreated soil, which strongly influences the overall behavior of the improved ground as a system. A further difference is that the treated soil in Nordic applications is considered as vertical drainage, whereas the Japanese treated soils are practically impermeable materials.

The major purpose of the Nordic applications is the reduction of settlement, and a group of columns is installed underneath road embankments or around dwellings. In comparison, the Japanese application is initiated to improve the stability of port facilities such as breakwaters and revetments in which the pattern of application is massive stabilization created in-situ by overlapping stabilized columns. The principle of the Deep Mixing Method in Nordic countries and in Japan is the same, but their applications vary in the two regions of the world.

The current text is the latest State of Practice, which was compiled from two sources; one is a technical manual originally published in 1999 in Japanese (Coastal Development Institute of Technology, 1999) and the other is a course material for the University of Hong Kong prepared in English (Terashi, Kitazume & Tsuboi, 1997). The descriptions in the following chapters of the current text are mostly based on the research done in Japan or on accumulated experience in Japan.

References

Asano, J., K. Ban, K. Azuma & K. Takahashi. 1996. Deep mixing method of soil stabilization using coal ash, Grouting and Deep Mixing. *Proc. of the 2nd International Conference on Ground Improvement Geosystems*, 1: 393-398. Balkema

Assarson, K.G. et al. 1974. Deep stabilization of soft cohesive soils. *Linden-Alimak*. Skellefteå.

Babasaki, R, M. Terashi, T. Suzuki, A. Maekawa, M. Kawamura & E. Fukazawa. 1996. Japanese Geotechnical Society Technical Committee Report - Factors influencing the strength of improved soil, Grouting and Deep Mixing. *Proc. of the 2nd International Conference on Ground Improvement Geosystems*, 2: 913-918. Balkema.

Broms, B.B. & P. Boman. 1975. Lime stabilized column. *Proc. of the 5th Asian Regional Conference on Soil Mechanics and Foundation Engineering*, 1: 227-234.

Broms, B. B. & P. Boman. 1977. Stabilization of soil with lime columns. *Royal Institute of Technology*, Stockholm.

Cement Deep Mixing Method Association. 1993. *Cement Deep Mixing Method Manual* (in Japanese).

Coastal Development Institute of Technology. 1999. *Technical Manual of Deep Mixing with Respect to Marine Works*. 226 p. (in Japanese).

Hayashi, H., S. Noto & N. Toritani. 1989. Cement improvement of Hokkaido peat. *Symposium on High Organic Soils*: 101-106 (in Japanese).

Ingles O.G. & J.B. Metcalf. 1972. *Soil stabilization, Principles and Practice.* Butterworth.

Japan Cement Association. 1994. *Soil improvement manual using cement stabilizer* (in Japanese).

Japanese Geotechnical Society. 2000. Practice for making and curing stabilized soil specimens without compaction. *JGS 0821-2000*. Japanese Geotechnical Society (in Japanese).

Ministry of Transport. 1980. *Study Needs on Development of Technologies related to the Port and Harbor Construction - Present and Outlook of Construction Technology* (in Japanese).

Ministry of Transport. 1986. Improvement of soft soils (II). *Internal Report of the Ministry of Transport* (in Japanese).

Noto, S., N. Kuchida & M. Terashi. 1983. Case histories of the deep mixing method. *Proc. of the Journal of Japanese Society of Soil Mechanics and Foundation Engineering, Tsuchi To Kiso*, 31(7):73-80 (in Japanese).

Ogawa, F. & K. Matsumoto. 1978. Correlation between engineering coefficients of soils in the port and harbour regions. *Report of the Port and Harbour Research Institute*, 17(3): 3 – 8 (in Japanese).

Okumura, T., T. Mitsumoto, M. Terashi, M. Sakai & T. Yoshida. 1972. Deep lime mixing method for soil stabilization (1st report). *Report of the Port and Harbour Research Institute*, 11(1): 67-106 (in Japanese).

Okumura, T. & M. Terashi. 1975. Deep lime mixing method of stabilization for marine clays. *Proc. of the 5th Asian Regional Conference on Soil Mechanics and Foundation Engineering*, 1: 69-75.

Public Works Research Center. 1999. *Technical Manual of Deep Mixing with Respect to on land applications*. 326 p. (in Japanese).

Rathmayer, H. 1996. Deep mixing method for soft subsoil improvement in the Nordic countries, Grouting and Deep Mixing. *Proc. of the 2nd International Conference on Ground Improvement Geosystems*, 2: 869-877. Balkema.

Saitoh, S., Y. Suzuki & K. Shirai. 1985. Hardening of soil improved by the deep mixing method. *Proc. of the 11th International Conference on Soil Mechanics and Foundation Engineering*, 3: 1745-1748.

Terashi, M. 1983. Problems and research orientation of the deep mixing method. *Proc. of the Journal of Japanese Society of Soil Mechanics and Foundation Engineering, Tsuchi To Kiso*, 31(8): 75-83 (in Japanese).

Terashi, M., H. Tanaka & T. Okumura. 1979. Engineering properties of lime treated marine soils and the DM method. *Proc. of the 6th Asian Regional Conference on Soil Mechanics and Foundation Engineering*, 1: 191-194.

Terashi, M. & H. Tanaka. 1981. Ground improved by the deep mixing method. *Proc. of the 10th International Conference on Soil Mechanics and Foundation Engineering*, 3: 777-780.

Terashi, M. & H. Tanaka. 1983. Settlement analysis for the deep mixing method. *Proc. of the 8th European Conference on Soil Mechanics and Foundation Engineering*, 2: 955-960.

Terashi, M., H. Fuseya & S. Noto. 1983a. Outline of the deep mixing method. *Proc. of the Journal of Japanese Society of Soil Mechanics and Foundation Engineering, Tsuchi To Kiso*, 31(6): 47-54 (in Japanese).

Terashi, M., H. Tanaka & M. Kitazume. 1983b. Extrusion failure of ground improved by the Deep Mixing Method. *Proc. of the 7th Asian Regional Conference on Soil Mechanics and Foundation Engineering*, 1: 313-318.

Terashi, M., H. Tanaka, T. Mitsumoto, S. Honma & T. Ohashi. 1983c. Fundamental properties of lime and cement treated soils (3rd report). *Report of the Port and Harbour Research Institute*, 22(1): 69-96 (in Japanese).

Terashi, M. 1997. Theme Lecture: Deep Mixing Method - Brief State of the Art. *Proc. of the 14th International Conference on Soil Mechanics and Foundation Engineering*, 4:2475-2478, Hamburg.

Terashi, M., M. Kitazume & H. Tsuboi. 1997. Development of the Deep Mixing Method and Current Practice. *Handout Material for Short Course*. The University of Hong Kong, November 27-29, 1997.

Watanabe S. 1974. Relations between soil properties. *Railway Technical Research Report*, 933:1-34 (in Japanese).

Yanase, S. 1968. Stabilization of marine clays by quicklime. *Report of the Port and Harbour Research Institute*, 7(4): 85-132 (in Japanese).

Yonekura R., M. Terashi & M. Shibazaki (eds). 1996. Grouting and Deep Mixing. *Proc. of the 2nd International Conference on Ground Improvement Geosystems*. Balkema.

CHAPTER 2

Factors Affecting Strength Increase

2.1 GENERAL

The magnitude of the strength increase of treated soil by lime or cement is influenced by a number of factors, because the basic strength increase mechanism is closely related to the chemical reaction between the soil and the stabilizing agent. The factors can be roughly divided into four categories: I. Characteristics of stabilizing agent, II. Characteristics and condition of soil, III. Mixing conditions, and IV. Curing conditions, as shown in Table 2.1 (Terashi, 1997).

Table 2.1. Factors affecting the strength increase (Terashi, 1997).

I.	Characteristics of stabilizing agent	1.	Type of stabilizing agent
		2.	Quality
		3.	Mixing water and additives
II.	Characteristics and conditions of soil (especially important for clays)	1.	Physical chemical and mineralogical properties of soil
		2.	Organic content
		3.	pH of pore water
		4.	Water content
III.	Mixing conditions	1.	Degree of mixing
		2.	Timing of mixing/re-mixing
		3.	Quantity of stabilizing agent
IV.	Curing conditions	1.	Temperature
		2.	Curing time
		3.	Humidity
		4.	Wetting and drying/freezing and thawing, etc.

Needless to say, the characteristics of stabilizing agent mentioned in Category I strongly affect the strength of treated soil. Therefore, the selection of an appropriate stabilizing agent is, in a real sense, an important issue. There are many types of stabilizing agent available in the Japanese market (Japan Cement

Association, 1994). The types of stabilizing agent and their stabilizing mechanisms have been described in the previous chapter. The factors in Category II (characteristics and conditions of soil) are inherent characteristics of each soil and the way that it has been deposited. It is usually impossible to change these conditions at the site to perform deep improvement. Thompson (1966) studied the influence of the properties of Illinois soils on the lime reactivity of compacted lime-soil mixture and concluded that the major influential factors were acidity (pH) and organic carbon content of the original soil. Japanese research groups have also performed similar studies on lime or cement treated soils manufactured without compaction (Okumura et al., 1974, Kawasaki et al., 1978, 1981 and 1984, Terashi et al., 1977, 1979, 1980 and 1983). Their valuable works provide engineers with good qualitative information. The factors in Category III (mixing condition) are easily altered and controlled onsite during an execution based on the judgment of the engineers responsible for the execution. The factors in Category IV (curing conditions) can be controlled easily in a laboratory study but cannot be controlled at a work site. For detailed descriptions, see Terashi et al. (1977, 1980).

The influences of various factors on the strength of lime and cement treated soils are shown in Sections 2.2 and 2.3 respectively. In the following descriptions, the unconfined compressive strength, qu of the treated soil is used as an index representing the stabilized effect. The test specimens for the unconfined compression tests are, in principle, prepared in the laboratory by the procedure standardized by the Japanese Geotechnical Society (Japanese Geotechnical Society, 2000) which is briefly described in Appendix A. The influence factors unique to actual work sites and their effects are discussed in Chapter 6 and Appendix B.

2.2 INFLUENCE OF VARIOUS FACTORS ON THE STRENGTH OF LIME TREATED SOIL

(1) *characteristics of stabilizing agent*. Figure 2.1 shows the influence of the property of quicklime on the treated soil strength, in which four types of quicklime (named as A to D) with different activities were tested (Okumura et al., 1974). The activity is an index to represent the rate of hydration reaction of quicklime. The quicklimes A and B are quicklime with high activity burned at a low temperature (about 1000 °C), while the limes C and D are quicklime with low activity burned at a higher temperature. As shown in Figure 2.1(a), the lime A has the highest activity among them while the lime D has the lowest. In Figure 2.1(b), w_i and aw represent the initial water content of original soil and the lime content, respectively. Lime content, aw is defined by a ratio of the dry weight of stabilizing agent to the dry weight of original soil. The figure shows that the unconfined compressive strength, qu of the lime treated soil is much influenced by the activity of quicklime. This phenomenon is most dominant in which the strength increases by the limes A and B are much larger than those by the limes C and D. This emphasizes that an

appropriate type of lime should be selected to obtain high strength in lime improvement.

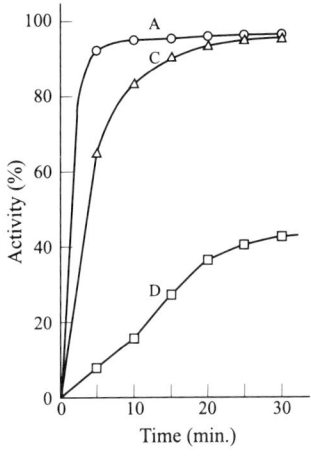
(a) activity of quicklimes

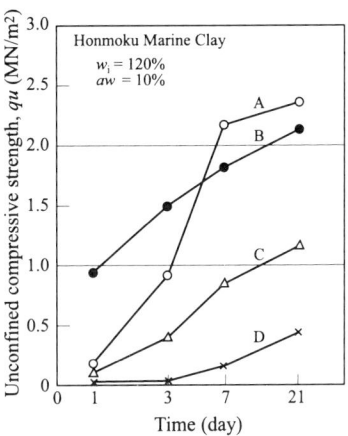
(b) relationship between strength increase and curing time

Figure 2.1. Influence of activity of quicklime on unconfined compressive strength (Okumura et al., 1974).

(2) characteristics and condition of original soil. Figure 2.2 shows the influence of the grain size distribution of original soil on the strength of quicklime treated soil (Terashi et al., 1977). In the test, Toyoura standard sand is added to two different clays so as to obtain artificial soils with different sand fractions. These artificial soils are stabilized with the same quicklime content, aw of 5% and 10%. In the figure, the unconfined compressive strength, qu after 7 days' curing, is shown on the vertical axis. It can be seen that the unconfined compressive strength is influenced by the amount of sand fraction and has a peak value at a sand fraction of around 40 to 60%.

20 *The Deep Mixing Method – Principle, Design and Construction*

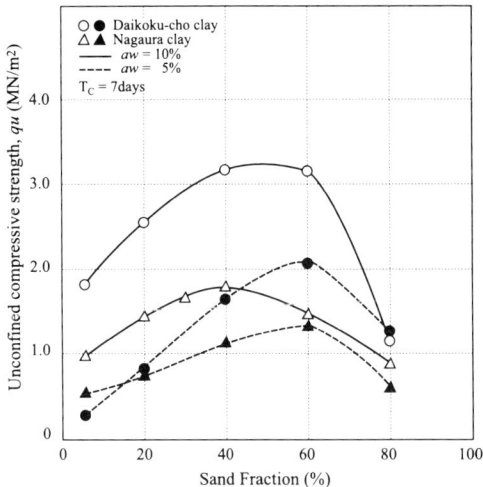

Figure 2.2. Influence of grain size distribution on strength (Terashi et al., 1977).

Figures 2.3 and 2.4 are test data obtained with the compacted lime treated Illinois soils (Thompson, 1966). These two figures directly or indirectly explain the influence of soil acidity on the strength. The influence of the pH of original soil on the unconfined compressive strength, q_u, is shown in Figure 2.3. Although there is a large scatter in the test data, the tendency of decreasing strength with decreasing pH is apparent. Figure 2.4 shows the relationship between the q_u and the organic carbon content of original soil. The strength increase becomes negligibly small when the organic carbon content of original soil exceeds about 1%.

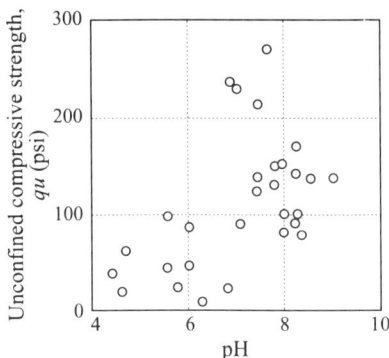

Figure 2.3. Influence of soil pH on strength of compacted lime treated soil (Thompson, 1966).

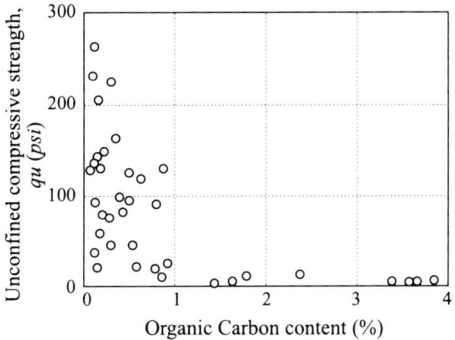

Figure 2.4. Influence of organic carbon content on strength of compacted lime treated soil (Thompson, 1966).

The influence of the water content of original soil on the unconfined compressive strength is shown in Figure 2.5 (Terashi et al., 1977). Honmoku marine clay prepared at different initial water contents, w_i, was stabilized by two different lime contents and cured for 3, 7 and 21 days until the unconfined compression test. The maximum effects are achieved at around the liquid limit of the original soil, w_l for short-term strength at Tc of 3 days. With increasing curing time the optimum water content shifts toward the dry side. The improvement effect considerably decreases with increasing the water content when it exceeds the liquid limit, w_l. In the case of marine construction in Japan, this phenomenon might not cause serious problems because the natural water content of normally consolidated Japanese marine clay is close to its liquid limit in most cases. Care, however, should be taken in sea reclamation areas with pump dredged clay.

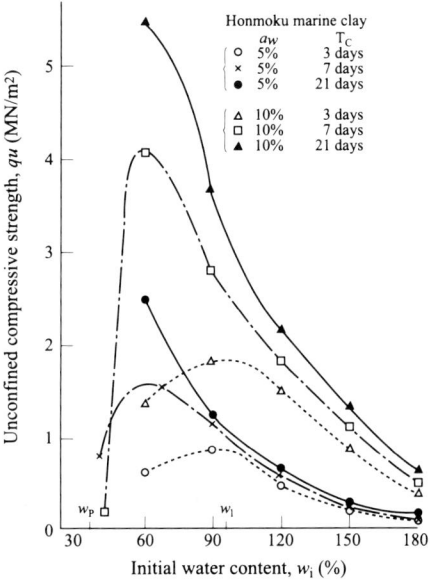

Figure 2.5. Influence of initial water content on strength (Terashi et al., 1977).

(3) *mixing conditions*. Terashi et al. (1977) investigated the influence of mixing time on the unconfined compressive strength, qu, by changing the mixing time of laboratory mixer in manufacturing laboratory specimens. In the tests, Kawasaki clay with various initial water contents, w_i, were treated with quicklime. The vertical axis of Figure 2.6 shows the strength ratio, which is defined by a ratio of strength of a treated soil prepared with an arbitrary mixing time to the strength of the soil with a mixing time of 10 minutes. The strength ratio decreases considerably when the mixing time is less than 10 minutes, especially for the case of a smaller lime content. When the mixing time exceeds 10 minutes, the strength ratio increases slightly with increasing the mixing time. A similar tendency was confirmed on cement treated soil by Nakamura et al. (1982), as is shown later in Figure 2.15.

In the above description, the mixing time is an index to represent how sufficiently the mixing of stabilizing agent and soil has been achieved. The degree of mixing depends not only on the mixing time but also on the type of mixer and the characteristics of original soil to be stabilized. Based on the past experience of Japanese alluvial clay with a water content around the liquid limit, Terashi et al. (1977) proposed a mixing time of 10 minutes and the use of a recommended soil mixer. In running the laboratory tests with different soils and a different mixer, a responsible engineer should go to the laboratory to confirm the most appropriate mixing time. The recent laboratory procedure standardized by the Japanese Geotechnical Society (Japanese Geotechnical Society, 2000) simply prescribes "sufficient mixing" in the main text and suggests 10 minutes in the supplements (see Appendix A).

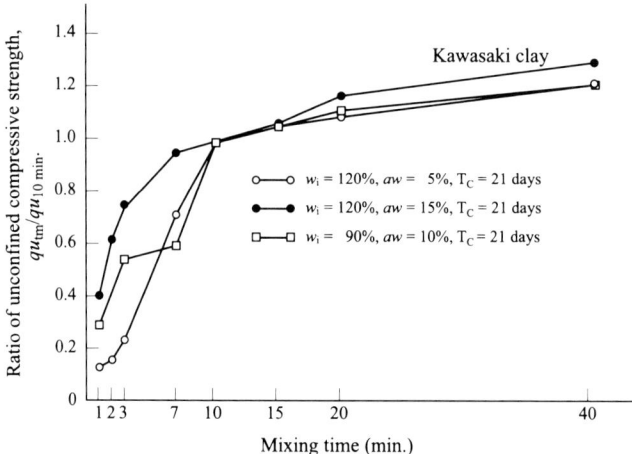

Figure 2.6. Influence of mixing time on strength (Terashi et al., 1977).

Figure 2.7 shows the relationship between the amount of quicklime, aw, and the unconfined compressive strength, qu, in which two different marine soils were tested (Terashi et al., 1977). In the case of Yokohama reclaimed soil (w_l of 78.8%

and w_p of 39.1%), the unconfined compressive strength increases almost linearly with increasing the amount of stabilizing agent, irrespective of the curing time. In the case of Honmoku marine clay (w_l of 92.3% and w_p of 46.9%) however, a clear peak strength can be seen and also the amount of agent at the peak strength becomes larger with a longer curing time.

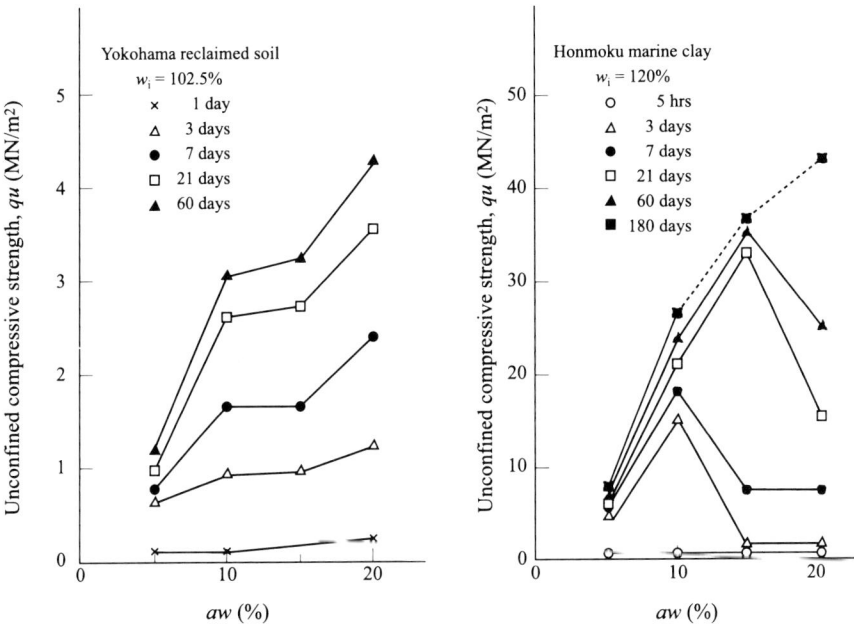

Figure 2.7. Influence of amount of agent in lime stabilization (Terashi et al., 1977).

(4) *curing conditions*. Figure 2.8 shows the influence of curing time on the unconfined compressive strength, qu, of various kinds of clay stabilized with the same lime content of 10% (Terashi et al., 1977). The strength increase is much dependent upon the clay type even when the amount of agent is the same, but the compressive strength of all the clays increases almost linearly with the logarithm of curing time. The strength increase of treated soils for more than 10 years will be shown later in Figure 3.12, in which the strength also increases almost linearly with the logarithm of curing time for longer term.

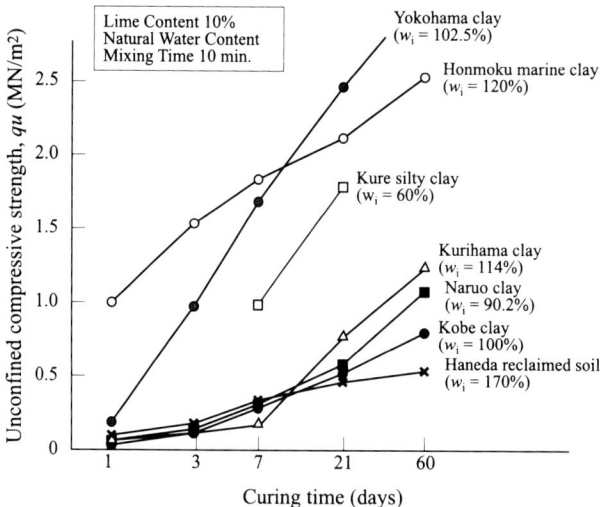

Figure 2.8. Influence of curing time in lime stabilization (Terashi et al., 1977).

2.3 INFLUENCE OF VARIOUS FACTORS ON THE STRENGTH OF CEMENT TREATED SOIL

(1) *characteristics of stabilizing agent*. Figure 2.9 shows the influence of different cements on the treated soil strength in which Portland cement and blast furnace slag cement type-B were compared at a curing time, Tc of 28 days to 5 years (Saitoh, 1988). The tests were conducted on two different sea bottom sediments; Yokohama Port clay and Osaka Port clay. For each clay three different amounts of cement were applied. The cement factor, α (kg/m^3), is defined as a dry weight of cement added to 1 m^3 of original soil. The horizontal axis of the figure shows the curing time, Tc. The vertical axis of the upper figures for each clay is the unconfined compressive strength, qu, of the treated soil. The vertical axis of the lower figures is the unconfined compressive strength at arbitrary curing time, Tc normalized by that of 28 days' strength: qu_{Tc}/qu_{28}. In the case of Yokohama Port clay which exhibits high pozzolanic reaction, Portland cement is much more effective than blast furnace slag cement type B. Whereas in the case of Osaka Port clay with smaller pozzolanic reaction, blast furnace slag cement type B is much more effective. These test results suggest that the appropriate selection of the type of cement may be made if the pozzolanic reaction of the soil is known beforehand. It is interesting to see the qu_{Tc}/qu_{28} is higher for blast furnace slag cement irrespective to the difference of soil type.

Factors Affecting Strength Increase 25

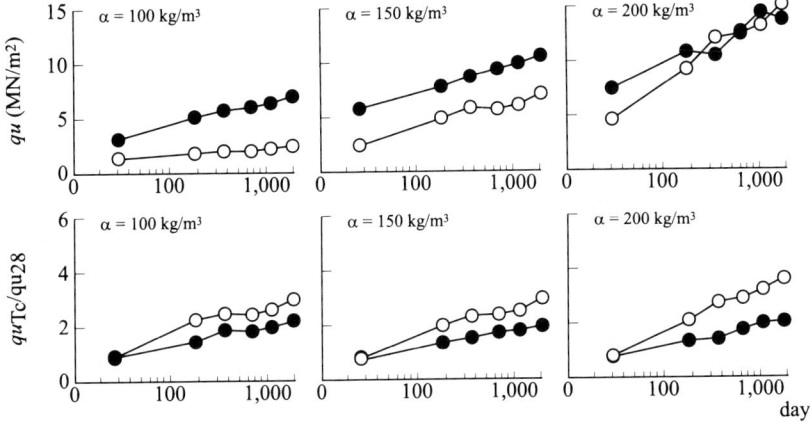

(a) Ground improved with clay from the Port of Yokohama

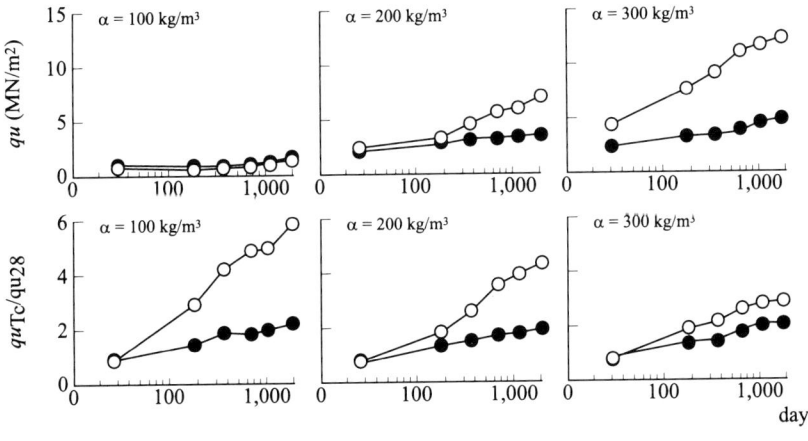

(b) Ground improved with clay from the Port of Osaka

● : Portland cement
○ : Blast-furnace slag cement type B.

Figure 2.9. Influence of cement type on unconfined compressive strength (Saitoh, 1988).

(2) *characteristics and conditions of soil*. The influence of soil type on the unconfined compressive strength, qu is shown in Figure 2.10, in which a total of 21 different soils were stabilized by Portland cement at a cement content, aw of 20% (Niina et al., 1981). In the figure, various physical and chemical properties of the original soils are also shown. The figure indicates that the humus content and pH of the original soil are the most dominant factors influencing the strength.

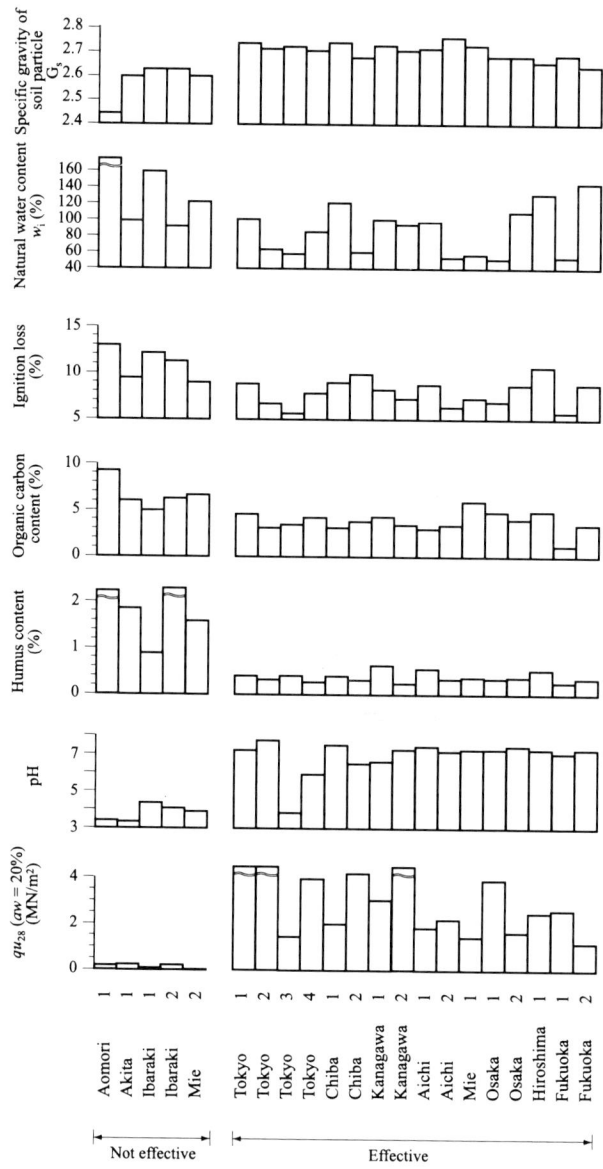

Figure 2.10. Influence of soil type in cement stabilization (Niina et al., 1981).

Figure 2.11 shows the influence of the grain size distribution on the unconfined compressive strength, qu of cement treated soil (Niina et al., 1977). Four artificial soils with different grain size distributions (Fig. 2.11(b)) were stabilized by Portland cement with three cement factors, α. In their tests, Ooigawa sand was added to Shinagawa alluvial clay to make various grain size distributions. Here soils B and C are prepared artificially by blending clayey soil A and sand D in order to investigate the influence of sand fraction. Unconfined compression tests were carried out after 28 days of curing. Similarly to the lime treated soil as already shown in Figure 2.2, the unconfined compressive strength, qu is dependent upon the sand fraction and the highest improvement effect can be achieved at a sand fraction of around 60% irrespective of the amount of cement. This amount of sand fraction is quite close to that for the lime treated soil (Fig. 2.2).

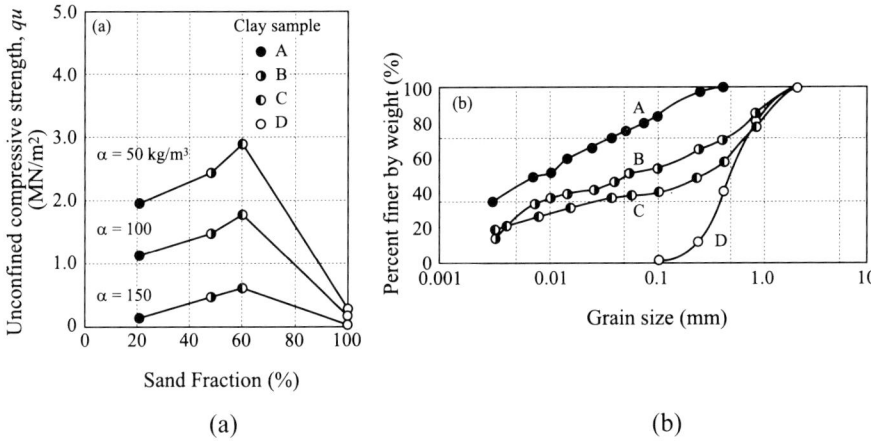

Figure 2.11. Influence of grain size distribution in cement stabilization (Niina et al., 1977).

Figure 2.12 shows the influence of the humin acid content of the original soil on the unconfined compressive strength (Miki et al., 1984). Artificial soil samples were prepared by adding various amount of humin acid to kaolin clay, in which humin acid of 0% to 5% of the dry weight of the kaolin clay was added. In the tests, these artificial soils were stabilized by nine kinds of stabilizing agent whose chemical compositions are shown in Figure 2.12(a). Figure 2.12(b) shows the relationship between the unconfined compressive strength, qu and the humin acid content. It is found that the unconfined compressive strength of the treated soil is much dependent upon the stabilizing agent, but decreases considerably with the increase of humin acid content irrespective of the type of stabilizing agent.

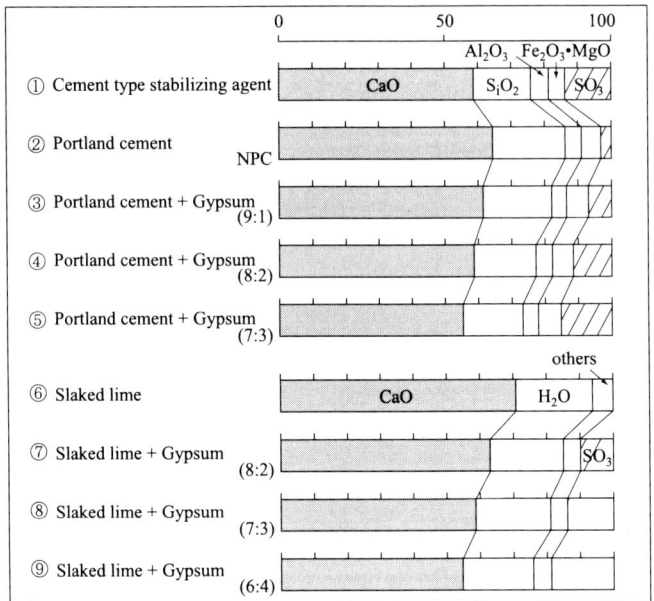

(a) Chemical composition of stabilizing agents

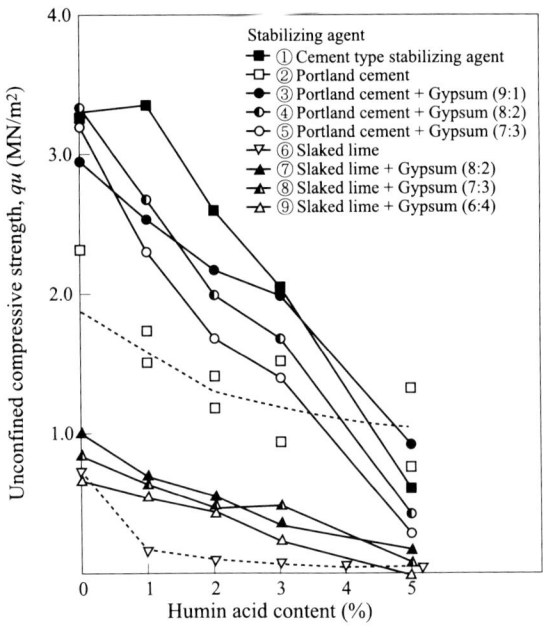

(b) Unconfined compressive strength, q_u

Figure 2.12. Influence of humin acid content on unconfined compressive strength (Miki et al., 1984).

Figure 2.13 shows the influence of pH on the unconfined compressive strength, qu. In the figure, the test results of five different soils are plotted, where major characteristics are tabulated in the attached table. On the horizontal axis, a new parameter F is plotted to incorporate the influence of pH, which is defined by the following equation (Nakamura et al., 1980).

$$\left.\begin{array}{ll} F = Wc / (9 - pH) & \text{for pH} < 8 \\ F = Wc & \text{for pH} > 8 \end{array}\right\} \quad \cdots\cdots\cdots\cdots\cdots \text{Eq. (2.1)}$$

in which
    Wc : dry weight of cement added to original soil of 1 m³

The qu value is roughly proportional to F and the relation between qu and F is found to be $qu = 32.5\ F - 1.625$ (MN/m²). However, this relation is not used in the prediction of the strength due to the large scatter in the test results.

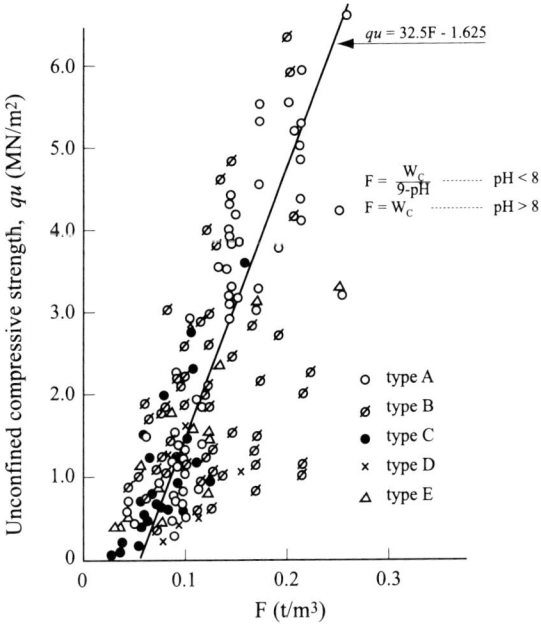

Type	Wet unit weight, γ_t (g/cm³)	Natural water content w_n (%)	Liquid limit w_l (%)	Plastic limit w_p (%)	Grain size distribution (%)			pH (H₂O)	Ignition loss
					Sand, gravel fraction	Silt fraction	Clay fraction		
A	1.38~1.76	55~144	51~121	2~18	30~47	33~50	29~52	8.1~8.7	2.0~7.0
B	1.28~1.70	38~160	27~204	3~42	18~76	27~70	16~54	7.3~8.9	3.8~12.9
C	1.50~1.76	42~86	49~110	3~43	22~66	36~54	12~55	5.5~7.9	4.0~12.0
D	1.10~1.40	114~740	–	–	–	–	–	5.5~6.0	19.0~64.0
E	1.49~1.97	25~56	–	49~86	–	9~35	2~22	5.4~9.3	3.3~15.2

Figure 2.13. Effects of organic carbon content (Ignition Loss) and pH in cement treated soil (Nakamura et al., 1980).

The influence of the initial water content of the original soil on the unconfined compressive strength, q_u, is shown in Figure 2.14 (CDIT, 1999). In the tests, two kinds of marine clays were stabilized by Portland cement and blast furnace slag cement type B. The unconfined compressive strength decreases almost linearly with increasing water content.

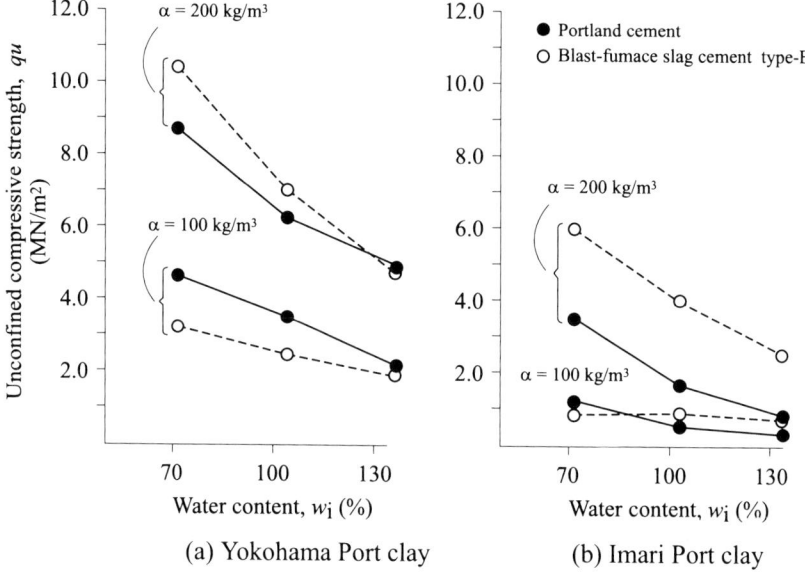

Figure 2.14. Influence of initial water content on strength (Tc of 91 days) (CDIT, 1999).

(3) *mixing conditions*. Figure 2.15 shows the relationship between the mixing time and the unconfined compressive strength, q_u in a laboratory mixing tests (Nakamura et al., 1982). The laboratory mixing tests were conducted as the same manner as the standardized procedure except the mixing time. In the tests, Portland cement is added to the original soil either in a dry form or a slurry form with a water and cement ratio, w/c of 100%. The unconfined compressive strength decreases with the decrease of mixing time. The figure also shows that the strength deviation increases with the decrease of mixing time.

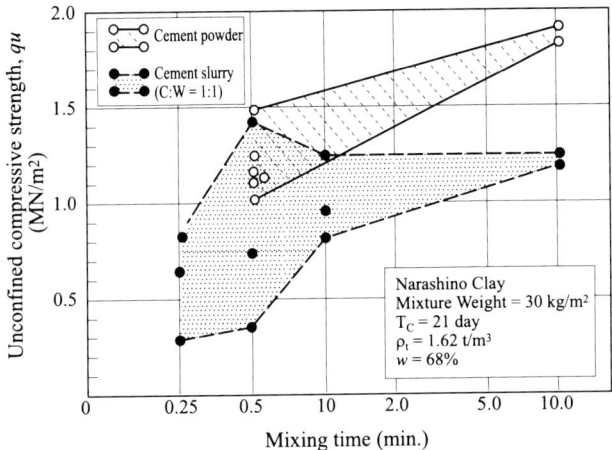

Figure 2.15. Influence of mixing time on strength deviation (Nakamura et al., 1982).

Figure 2.16 shows the influence of the amount of cement, aw, on the unconfined compressive strength, qu, in which Kawasaki clay with the initial water content of 120% was stabilized by Portland cement with various amounts of cement, and was tested after four various curing periods. The unconfined compressive strength increases almost linearly with the increasing amount of cement, as shown in Figure 2.16. The figure also shows that a minimum amount of cement of about 5% is necessary irrespective of the curing period to obtain improvement effect for this particular soil.

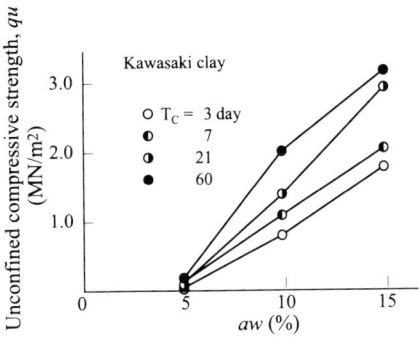

Figure 2.16. Influence of amount of cement on strength (Terashi et al., 1980).

A similar phenomenon for organic soils is shown in Figure 2.17, in which the weight of cement per 1 m^3 of the original soil is plotted on the horizontal axis (Babasaki et al., 1980). The strength achieved is relatively small in the organic soils, but it increases with increasing cement content. The minimum cement content for these organic soils is around 50 kg/m^3.

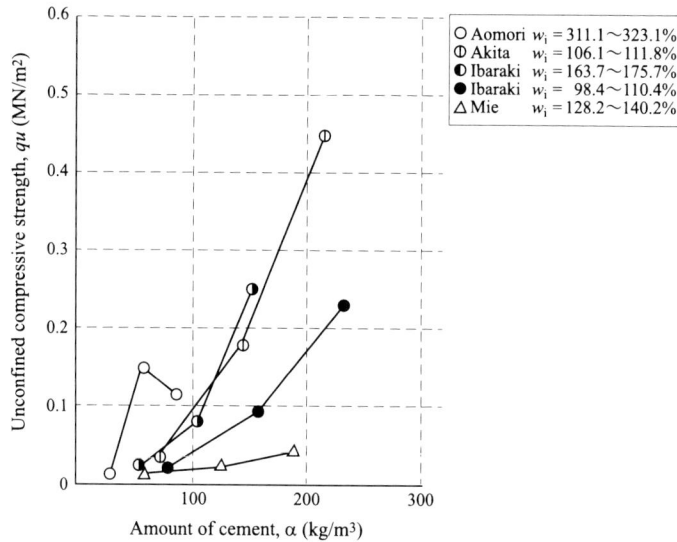

Figure 2.17. Influence of amount of cement on strength for organic soils (Babasaki et al., 1980).

(4) *curing conditions*. The influence of curing temperature is shown in Figure 2.18, in which the treated soils (Yokohama and Osaka clays) were cured at different temperatures for up to four weeks (Saitoh et al., 1980). In the figure, the strength of treated soil cured at an arbitrary temperature is normalized by the strength of treated soil cured at a temperature of 20 degrees Celsius. The figure shows that a higher strength can be obtained under a higher curing temperature. This influence of the curing temperature is more dominant for short-term strength but it diminishes as the curing time becomes longer.

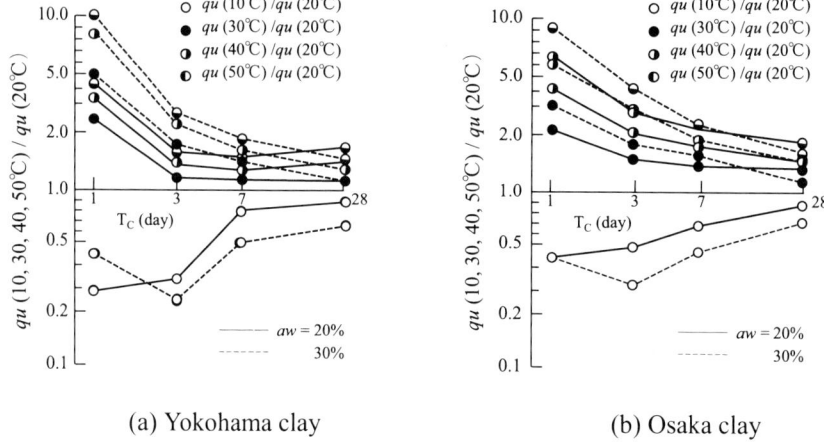

(a) Yokohama clay   (b) Osaka clay

Figure 2.18. Effects of curing temperature (Saitoh et al., 1980).

Figure 2.19 shows the strength increase of cement treated soil with the curing time (Kawasaki et al., 1981). The unconfined compressive strength, qu increases irrespective of the soil type, and the strength increase with time is more dominant for a larger amount of agent. Similar test results were obtained with Portland cement or fly ash cement (Saitoh, 1988).

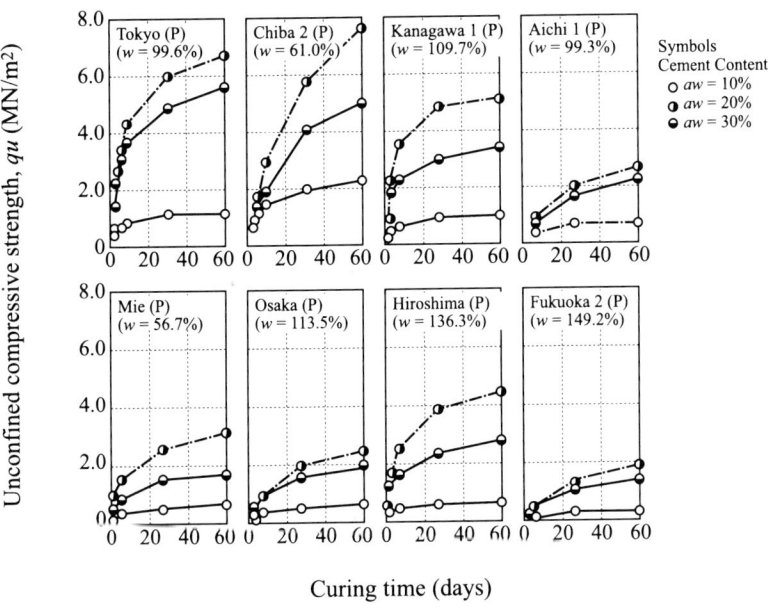

Figure 2.19. Strength increase with curing time (Kawasaki et al., 1981).

2.4 PREDICTION OF STRENGTH

In the process of soil improvement work, the strength of the in-situ treated soil should be predicted at each stage of the planning, testing, design, and implementation. There are a lot of proposed formulas to predict the laboratory strength and field strength, which incorporate the various factors for the improvement effect. The general formula may be written as:

qu_l = function (soil type, stabilizing agent, C/Wt, Oc, Fc, Tc, etc.) ·· Eq. (2.2)

qu_f = function (qu_l, Tc, θ, mixedness, environment) ············ Eq. (2.3)

where
- C/Wt: ratio of the weight of the stabilizing agent to that of total water including mixing water
- Fc: fine grain content (may be substituted by the amount of soluble silica and alumina)
- Oc: organic matter content (may be substituted by pH or Ig. loss)

qu_f: field strength of the soil manufactured in-situ with differing mixing and curing conditions

qu_l: strength of the soil prepared in the laboratory by sufficient mixing and cured under standardized curing conditions

Tc: curing time

θ : curing temperature

soil type: characteristics of original soil

stabilizing agent: type and quality of stabilizing agent

mixedness: degree of mixing

environment: humidity, manufacturing conditions, etc.

Many papers have proposed a simplified version of the above formula for predicting qu_l and making comparisons with laboratory test results. However, we are not yet at the stage where we can predict the laboratory strength with a reasonable level of accuracy. Furthermore, there is no widely applicable formula for estimating the field strength which incorporates all the relevant factors, because the strength of in-situ treated soil is influenced by the mixing and curing conditions, which differ from one machine to another and according to specific site conditions. Because of this, most predictions are now made by performing laboratory mold tests and then estimating the field strength on the basis of past experience. In large scale projects, test results are often checked by field trial tests, and for small scale work, reference is made to previous cases in similar areas. There are also cases where these sorts of test result are used to derive a formula that can be applicable only to the particular site.

References

Babasaki, R., T. Kawasaki, A. Niina, R. Munechika & S. Saitoh. 1980. Study of the deep mixing method using cement hardening agent (Part 9). *Proc. of the 15th Japan National Conference on Soil Mechanics and Foundation Engineering*: 713-716 (in Japanese).

Coastal Development Institute of Technology. 1999. *Deep Mixing Method Technical Manual for Marine Works*: 147 (in Japanese).

Japan Cement Association. 1994. *Soil Improvement Manual Using Cement Stabilizer*: 424 (in Japanese).

Japanese Geotechnical Society. 2000. Practice for making and curing non-compacted stabilized soil specimens. *JGS T 821-1990* (in Japanese).

Kawasaki, T., A. Niina, S. Saitoh & R. Babasaki. 1978. Studies on Engineering Characteristics of cement-base stabilized soil. *Takenaka Technical Research Report*, 19: 144 - 165 (in Japanese).

Kawasaki, T., A. Niina, S. Saitoh, Y. Suzuki & Y. Honjyo. 1981. Deep mixing method using cement hardening agent. *Proc. of the 10th Internal Conference on Soil Mechanics and Foundation Engineering*, 3: 721-724.

Kawasaki, T. et al. 1984. Deep mixing method using cement slurry as hardening agent. *Seminar on Soil Improvement and Construction Techniques in soft Ground, Singapore.*

Miki, H., K. Kudara & Y. Okada. 1984. Influence of humin acid content on ground improvement (part 2). *Proc. of the 39th Annual Conference of the Japan Society of Civil Engineers*, 3: 307-308 (in Japanese).

Nakamura. M, H. Akutsu & F. Sudo. 1980. Study of improved strength based on the deep mixing method (Report 1). *Proc. of the 15th Japan National Conference on Soil Mechanics and Foundation Engineering*: 1773 - 1776, (in Japanese).

Nakamura. M., S. Matsuzawa & M. Matsushita. 1982. Study of the agitation mixing of improvement agents. *Proc. of the 17th Japan National Conference on Soil Mechanics and Foundation Engineering*, 2: 2585-2588 (in Japanese).

Niina, A., S. Saitoh, R. Babasaki, I. Tsutsumi & T. Kawasaki. 1977. Study on DMM using cement hardening agent (Part 1). *Proc. of the 12th Japan National Conference on Soil Mechanics and Foundation Engineering*: 1325-1328 (in Japanese).

Niina, A., S. Saitoh, R. Babasaki, T. Miyata & K. Tanaka. 1981. Engineering properties of improved soil obtained by stabilizing alluvial clay from various regions with cement slurry. *Takenaka Technical Research Report*, 25: 1-21 (in Japanese).

Okumura, T., M. Terashi, T. Mitsumoto, T. Yoshida & M. Watanabe. 1974. Deep-lime – mixing method for soil stabilization (3rd Report). *Report of the Port and Harbour Research Institute*, 13, 2: 3-44 (in Japanese).

Saitoh, S., A. Niina & R. Babasaki. 1980. Effect of curing temperature on the strength of treated soils and consideration on measurement of elastic modules. *Proc. of Symposium on Testing of treated Soils,* Japanese Society of Soil Mechanics and Foundation Engineering: 61-66 (in Japanese).

Saitoh, S. 1988. *Experimental study of engineering properties of cement improved ground by the deep mixing method*. Ph.D. Thesis, Nihon University (in Japanese).

Terashi, M., T. Okumura & T. Mitsumoto. 1977. Fundamental properties of lime treated soil (1st report). *Report of the Port and Harbour Research Institute*, 16(1): 3-28 (in Japanese).

Terashi, M., H. Tanaka, T. Mitsumoto, Y. Niidome & S, Honma. 1979. Engineering properties of lime treated marina soils. *Proc. of the 6th Asian Regional Conference on soil Mechanics and Foundation Engineering*, 1: 191-194.

Terashi, M., H. Tanaka, T. Mitsumoto, Y. Niidome & S. Honma. 1980. Fundamental Properties of lime and cement treated soils (2nd report). *Report of the Port and Harbour Research Institute*, 19(1): 33-62 (in Japanese).

Terashi, M., H. Tanaka, T. Mitsumoto, S. Honma & T. Ohhashi. 1983. Fundamental properties of lime and cement treated soils (3rd Report). *Report of the Port and Harbour Research Institute*, 22(1): 69-96 (in Japanese).

Terashi, M. 1997. Theme Lecture: Deep Mixing Method - Brief State of the Art. *Proc. of the 14th International Conference on Soil Mechanics and Foundation Engineering*, 4: 2475-2478.

Thompson, R. 1966. Lime reactivity of Illinois soils. *Proc. of American Society of Civil Engineering*, 92, SM-5.

CHAPTER 3

Engineering Properties of Treated Soils

3.1 INTRODUCTION

The engineering properties of lime or cement treated soils have been extensively studied by highway engineers since the 1960s. However, the purpose of their treatment was to improve sub-base or sub-grade materials and the treatment was characterized by the low water content of original soils and small amounts of stabilizing agents. Mixing a few percent of stabilizing agent with respect to the dry weight of soil is enough to change the physical properties of a soil in order to enable efficient compaction that follows the mixing. In the case of deep improvement by DMM in Japan, a soft alluvial soil usually has natural water content nearly equal to or exceeding the liquid limit (w_l). Compaction of a soil-stabilizing agent mixture is impossible to carry out in depth. Due to these differences in manufacturing and in the expected function of treated soils, the fundamental engineering properties of lime or cement treated Japanese marine clays have been studied in detail. The major characteristics of lime or cement treated soils obtained in the laboratory and field are described briefly in the following paragraphs. Details of other characteristics such as dynamic properties, fatigue strength, creep etc. are found in Terashi et al. (1980, 1983) and in Kawasaki et al. (1981).

3.2 PHYSICAL PROPERTIES

(1) *change of water content*. Water content is altered by treatment due to chemical reaction, as described in Chapter 1. The water content changes of in-situ soils treated with quicklime are shown in Figure 3.1 (Kamata & Akutsu, 1976). In the field tests, 8 types of clay were stabilized by quicklime with lime content, aw of 10% to 25%. In the figure, estimated water content derived by the chemical reaction is also shown. It can be seen that the measured data almost coincides with the estimation. The water content profiles before and after cement treatments are plotted along the depth in Figure 3.2 (Kawasaki et al., 1978). In the field tests,

Tokyo Port clay (w_l of 93.1% and w_p of 35.8%) was stabilized by Portland cement with cement factor, α of 100 kg/m³ and 135 kg/m³ and with a water and cement ratio, w/c of 0.6. Although there is a large scatter along the depth, it can be seen that the water contents after treatment decrease about 20% from the original.

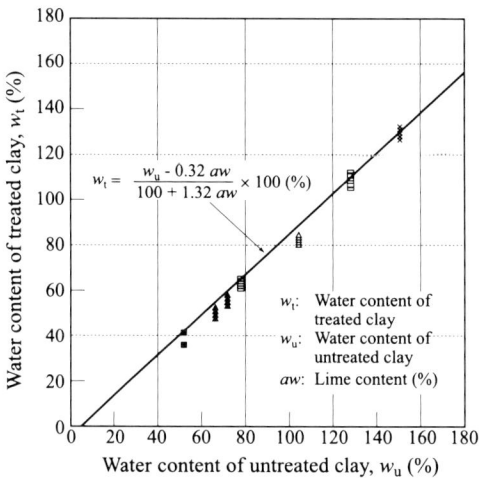

Figure 3.1. Change of water content by in-situ quicklime treatment (Kamata & Akutsu, 1976).

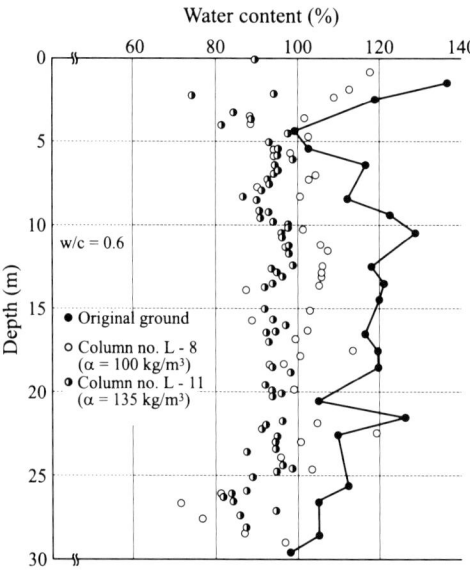

Figure 3.2. Change of water content by in-situ cement treatment (Kawasaki et al., 1978).

(2) *change of density*. Figure 3.3 shows the density change due to lime treatment (Kamata & Akutsu, 1976). The increase in density by treatment is relatively small, even if the water content is decreased, as already shown in Figure 3.1. The density change due to cement treatment is plotted against the amount of stabilizing agent in Figure 3.4 (Japan Cement Association, 1994). The vertical axis of the figure shows the wet density ratio of treated soil and untreated soil, ρ_t/ρ_{tu}. Although there is a large scatter in the test data, the wet density of the treated soil, ρ_t, increases about 3% to 15% due to cement treatment with dry form. For the cement treatment with slurry form, however, it is found that the density change due to treatment is negligible irrespective of water and cement ratio, w/c.

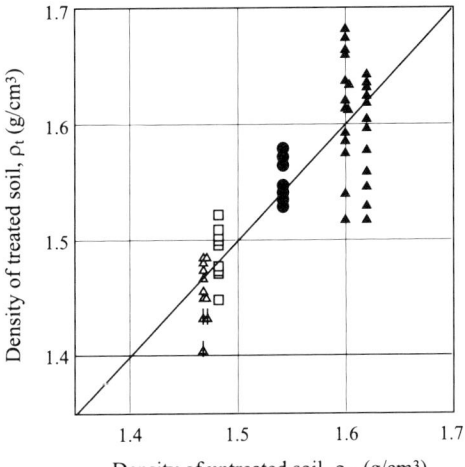

Figure 3.3. Change of density by in-situ quicklime treatment (Kamata & Akutsu, 1976).

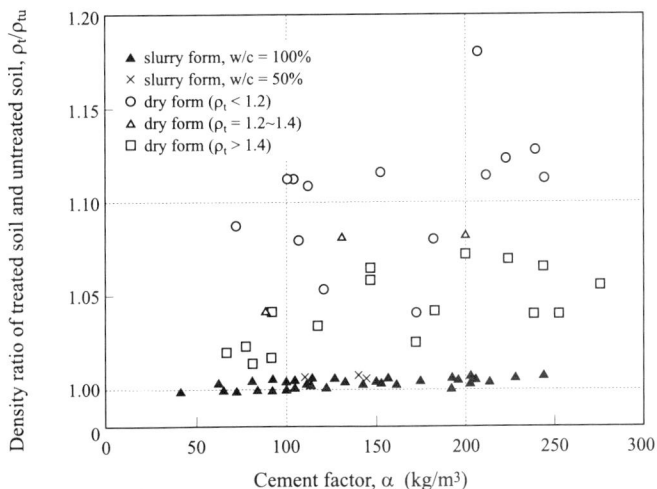

Figure 3.4. Change of density by in-situ cement treatment (Japan Cement Association, 1994).

3.3 MECHANICAL PROPERTIES (STRENGTH CHARACTERISTICS)

Based on the accumulated test data in Japan, no significant difference is found between the engineering characteristics of lime and cement treated soils. Therefore the mechanical properties of treated soil are described in the following sections without distinction of lime and cement treatment.

(1) stress - strain curve. A typical example of stress-strain curve of in-situ treated soil is shown in Figure 3.5, in which Tokyo Port clay (w_l of 93.1% and w_p of 35.8%) was stabilized by Portland cement with cement factor, α of 112 kg/m³ (Sugiyama et al., 1980). In the figure, the stress-strain curve of the original clay was also plotted. It is found that the stress-strain curve of the treated soil is characterized by very high strength and small axial strain at failure, while the original soil is characterized by small strength and large strain at failure.

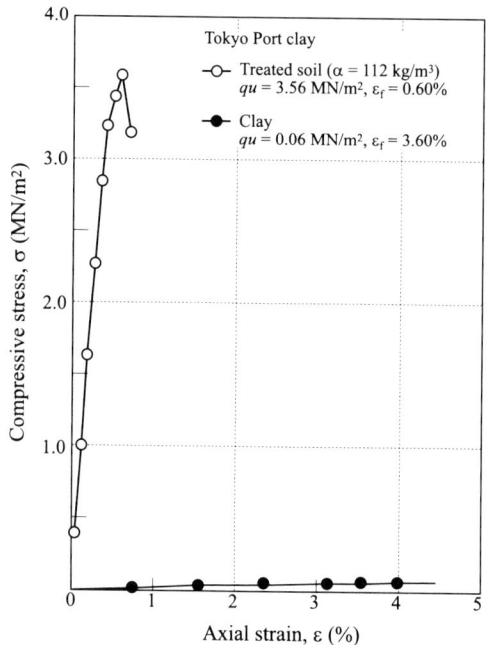

Figure 3.5. Stress - strain of in-situ treated soil (Sugiyama et al., 1980).

(2) strain at failure. Figure 3.6 shows the relationship between the axial strain at failure, ε_f, and the unconfined compressive strength, qu, of treated soils (Terashi et al., 1980). In the tests, Kawasaki clay (w_l of 87.7% and w_p of 39.7%) and Kurihama clay (w_l of 70.9% and w_p of 30.8%) were stabilized by either slaked lime, quicklime or Portland cement in the laboratory. In the undrained shear of treated soils, the magnitude of axial strain at failure is of the order of a few percent and markedly smaller than that of untreated marine clay. The axial strain at failure decreases with increasing the unconfined compressive strength, qu.

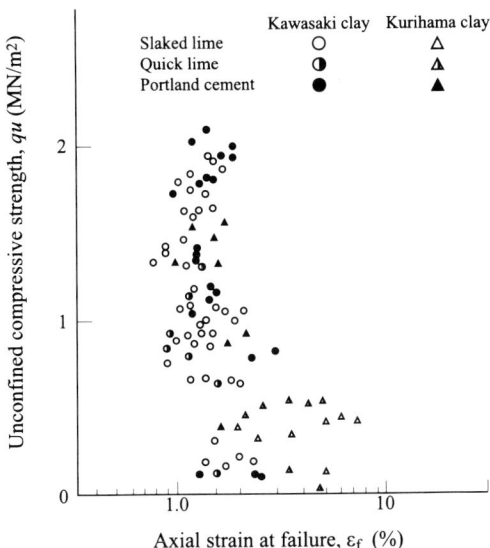

Figure 3.6. Strain at failure of treated soil in laboratory (Terashi et al., 1980).

(3) *modulus of elasticity*. The Young's modulus of elasticity of treated soils is plotted in Figure 3.7 against the unconfined compressive strength, qu (Terashi et al., 1977). The test data plotted in the figure are obtained on the specimen stabilized by quicklime in a laboratory. In the vertical axis of the figure, the modulus of elasticity, E_{50}, is plotted, which is defined by the secant modulus of elasticity in a stress - strain curve at a half of unconfined compressive strength. It can be seen in the figure that the modulus of the treated soils is 75 to 200 × qu when qu is less than 1.5 MN/m² and 200 to 1000 × qu when qu exceeds 1.5 MN/m². A similar relationship was obtained for the cement treated soil as shown in Figure 3.8, in which the E_{50} of cement treated soil is 350 to 1,000 × qu (Saitoh, 1985).

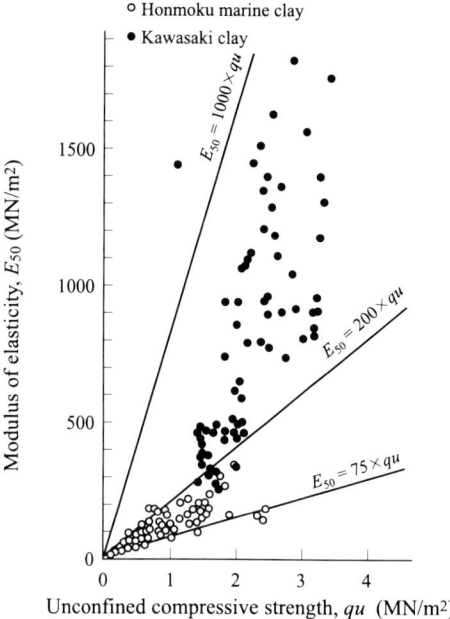

Figure 3.7. Modulus of elasticity, E_{50}, of lime treated soils stabilized in laboratory (Terashi et al., 1977).

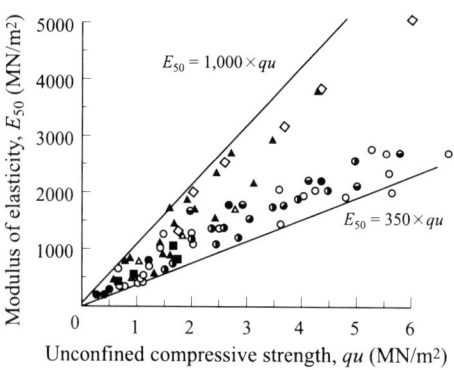

Figure 3.8. Modulus of elasticity, E_{50}, of cement treated soils stabilized in laboratory (Saitoh, 1985).

(4) *residual strength*. The residual strength of treated soil is almost zero in the case of unconfined compression. But even under small confined pressure, the residual strength of treated soil is increased to almost 80% of the unconfined compressive strength, qu (Tatsuoka & Kobayashi, 1983).

(5) *Poisson's ratio*. The Poisson's ratio, μ obtained of in-situ cement treated soils is shown in Figure 3.9 against the unconfined compressive strength, qu, in which

the unconfined compression tests were carried out on small scale specimens of 5 cm in diameter (Niina et al., 1977) and a large scale specimen of 100 cm in diameter (Hirade et al., 1995). Although there is a relatively large scatter in the test data, it can be seen that the Poisson's ratio of the treated soil is around 0.25 - 0.45, irrespective of the unconfined compressive strength, qu.

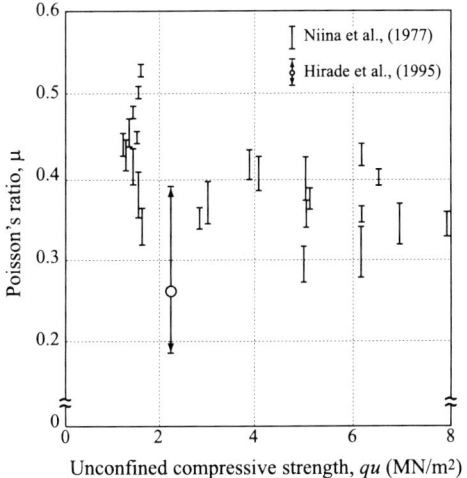

Figure 3.9. Poisson's ratio of in-situ cement treated soil (Niina et al., 1977 and Hirade et al., 1995).

(6) *angle of internal friction*. The previous researches show that the shear strength, cu of treated soil obtained in the unconsolidated – undrained compression test is almost constant irrespective of the confined stress, and the internal friction angle, Φ_u, is almost zero.

(7) *undrained shear strength*. The undrained shear strength, cu, obtained by isotropicaly-consolidated undrained compression test (CIU test), is almost constant as long as the consolidation pressure does not exceed the consolidation yield pressure (the pseudo pre-consolidation pressure), p_y. The undrained shear strength, cu, increases with increasing the consolidation pressure when the consolidation pressure exceeds the consolidation yield pressure, p_y, and the magnitude of cu is equivalent to that of the original clays consolidated to the same stress level.

(8) *bending strength*. The bending strength, σ_b, of treated soil was obtained from laboratory manufactured rectangular column specimen. Terashi et al performed a series of bending strength tests on Kawasaki clay stabilized either by quicklime or Portland cement. Their test results are plotted in Figure 3.10 against the unconfined compressive strength, qu. The bending strength of the treated soil is around 0.1- 0.6 of the unconfined compressive strength, qu, irrespective of the

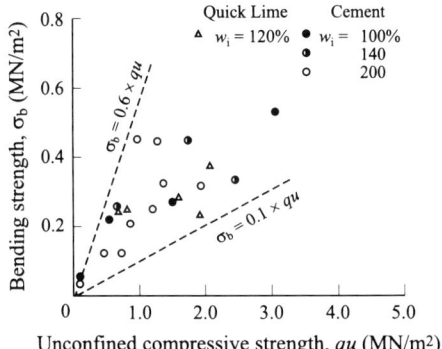

Figure 3.10. Bending strength of laboratory treated soils (Terashi et al., 1980).

type of stabilizing agent and the initial water content.

(9) *tensile strength*. The tensile strength of the treated soil manufactured in the laboratory is obtained by Brazilian tests (indirect tensile test). Figure 3.11 shows the relationship between the tensile strength, σ_t, and the unconfined compressive strength, qu. The figure shows the tensile strength, σ_t, increases almost linearly with increasing the unconfined compressive strength, qu, but shows a maximum value of 200 kN/m^2, irrespective of the type of stabilizing agent and the initial water content. The tensile strength is about 0.15 of the unconfined compressive strength, qu.

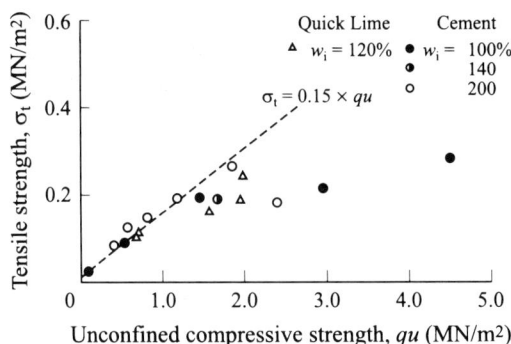

Figure 3.11. Tensile strength of laboratory treated soils (Terashi et al., 1980).

(10) *long-term strength*. The long-term strength of treated soils is shown in Figure 3.12. In the figure three treated soils stabilized in the field either by quicklime or Portland cement are plotted (CDIT, 1999, Terashi & Kitazume, 1992 and Yoshida et al., 1992). The figure shows the unconfined compressive strength of all the treated soils increase almost linearly with the increasing logarithm of curing time, irrespective of the type of clay, the type and amount of the stabilizing agent.

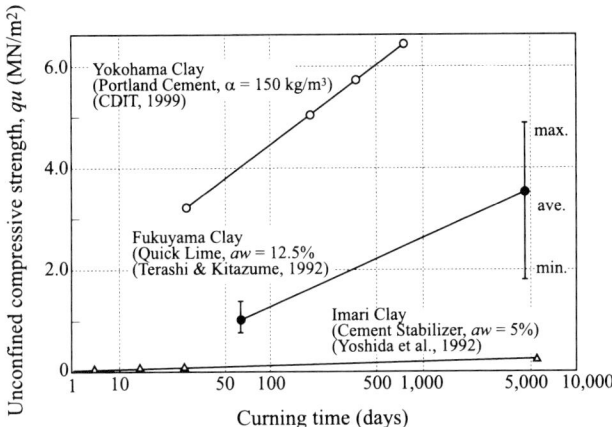

Figure 3.12. Long- term strength of in-situ treated soils (CDIT, 1999, Terashi & Kitazume 1992 and Yoshida et al., 1992).

3.4 MECHANICAL PROPERTIES (CONSOLIDATION CHARACTERISTICS)

(1) *consolidation yield pressure*. Figure 3.13 shows the void ratio and consolidation pressure curves (e - log p curves) of the laboratory cement treated soils, in which Tokyo Port clay (w_l of 93.1% and w_p of 35.8%) was stabilized by Portland cement with two different cement factors, α of 70 kg/m^3 and 100 kg/m^3 and cured 180 days (Kawasaki et al., 1978). In the laboratory consolidation tests, the treated soil samples with 2 cm and 6 cm in height and diameter respectively were consolidated one dimensionally up to 10 MN/m^2. The figure reveals that the consolidation phenomenon of the treated soils is similar to the ordinary clay samples, which is characterized by a sharp bend at a consolidation pressure. The consolidation pressure at the sharp bend is defined as a consolidation yield pressure, p_y. Kawasaki et al. (1978) also conducted similar consolidation tests on the in-situ stabilized soils and obtained a similar consolidation phenomenon.

Figure 3.14 shows the relationship between the consolidation yield pressure, p_y and the unconfined compressive strength, q_u, on Kawasaki clay and Kurihama clay stabilized by three different stabilizing agents. The figure shows that the consolidation yield pressure, p_y, has a linear relationship with the unconfined compressive strength, q_u. The ratio of p_y/q_u on the treated soils is approximately 1.3 irrespective of the type of original soil, and the type of stabilizing agent.

46 *The Deep Mixing Method – Principle, Design and Construction*

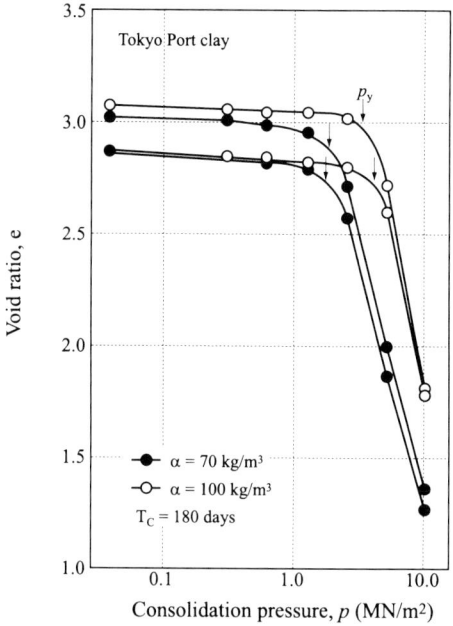

Figure 3.13. e - log p curves of laboratory treated soils (Kawasaki et al., 1978).

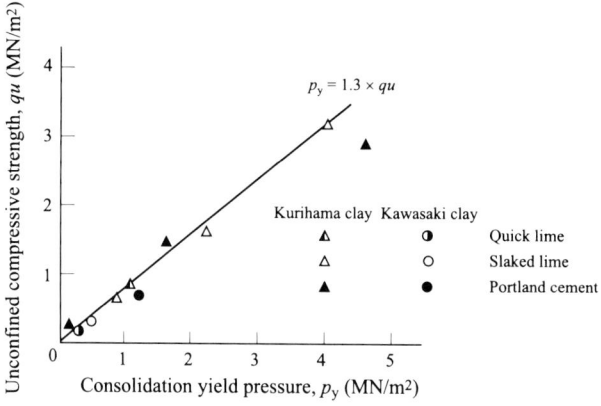

Figure 3.14. Consolidation yield pressure - unconfined compressive strength of laboratory treated soils (Terashi et al., 1980).

(2) *coefficient of consolidation and coefficient of volume co mpressibility*. Terashi et al. (1980) performed a series of conventional oedometer consolidation tests on the two marine clays stabilized by slaked lime, quicklime or Portland cement, in which the treated soil samples with 2 cm and 6 cm in height and diameter respectively were consolidated one dimensionally. Figure 3.15 shows the

relationship between the coefficient of consolidation of the treated clays, C_{vt} and the consolidation pressure, p. In the figure, the coefficient of consolidation of the treated soil, C_{vt}, normalized by that of the untreated soil, and the consolidation pressure, p, normalized by the consolidation yield pressure, p_y, are plotted on the vertical and horizontal axes, respectively. The figure shows the ratio of C_{vt}/C_{vu} is 10 to 100 as long as the normalized consolidation pressure, p/p_y, is around 0.1, in a sort of overconsolidated condition, but the C_{vt}/C_{vu} approaches to unity when the p/p_y exceeds 1, in a sort of normally consolidaed condition.

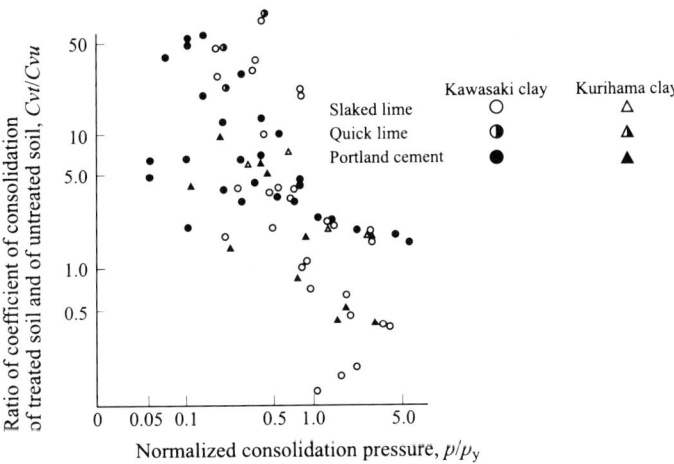

Figure 3.15. Relationship between coefficient of consolidation and consolidation pressure on laboratory treated soils (Terashi et al., 1980).

Figure 3.16 shows the relationship between the coefficient of volume compressibility of the treated clays, m_{vt}, and the consolidation pressure, p, as a similar manner to Figure 3.15. The figure shows the ratio of m_{vt}/m_{vu} is 1/10 to 1/100 as long as the normalized consolidation pressure, p/p_y, is around 0.1, in a sort of overconsolidated condition, but the m_{vt}/m_{vu} approaches to unity when the p/p_y exceeds 1, in a sort of normally consolidated condition. These figures indicate that the rate of consolidation of the treated soil is accelerated and the compressibility of the soil is reduced by lime and cement treatment as long as the consolidation pressure is lower than the consolidation yield pressure.

48 *The Deep Mixing Method – Principle, Design and Construction*

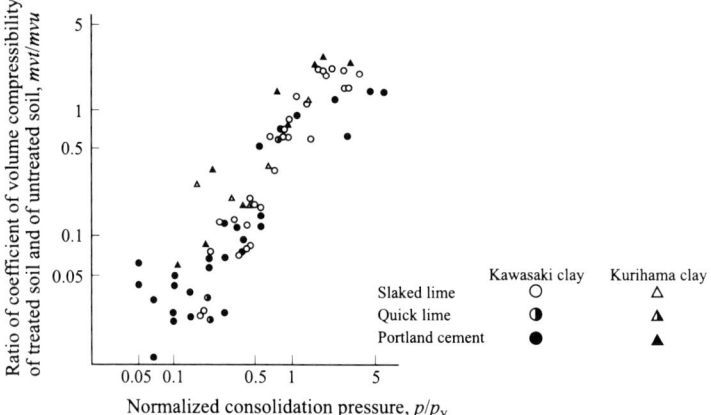

Figure 3.16. Relationship between coefficient of volume compressibility and consolidation pressure on laboratory treated soils (Terashi et al., 1980).

(3) *coefficient of permeability*. The coefficient of permeability of treated marine clays, k, is measured by constant head permeability tests in which the treated soils with 2 cm in height and 5 cm in diameter were tested (Terashi et al., 1983). Figure 3.17 shows the test result in which the coefficient of permeability is plotted against the water content of the treated soil. It is found that the permeability is dependent upon the water content of treated soil and amount of cement. The permeability of the treated soil decreases with decreasing the water content and with increasing the amount of cement. From accumulated test data on Japanese clays, it is known that the permeability of treated clays is equivalent to or lower than that of the untreated soft clays and whose order is 10^{-9} to 10^{-6} cm/sec (Fig. 3.17). Therefore in Japan the treated soil is not expected to function as a drainage layer.

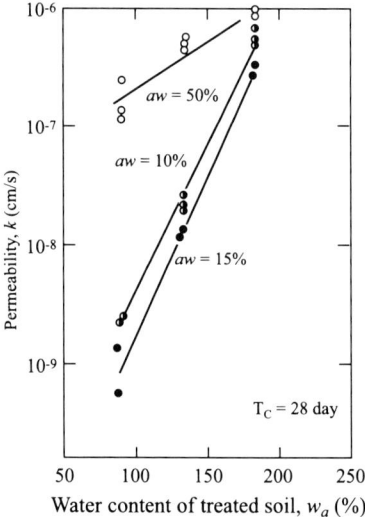

Figure 3.17. Relationship between permeability and water content of cement treated soils (Terashi et al., 1983).

3.5 ENGINEERING PROPERTIES OF CEMENT TREATED SOIL MANUFACTURED IN-SITU

(1) mixedness and stabilizing agent content in-situ. The engineering properties of the treated soils mentioned in the previous section are obtained mostly on the laboratory treated soil specimens with sufficient mixedness. In actual practice, original soil and stabilizing agent are mixed by a DM machine in-situ with a lower degree of mixedness in comparison with the laboratory specimens. If both the mixedness and the stabilizing agent content are low, the uniform mixing of the original soil and stabilizing agent cannot be attained. The amount of stabilizing agent often applied in actual practice in Japan is 10 to 20% of the dry weight of original soil. Thus, the average of unconfined compressive strength, qu, of in-situ treated soils easily exceeds 1 MN/m² in the case of reactive soils.

(2) unconfined compressive strengths of treated soils manufactured in-situ and in laboratory. Because of lower mixedness in-situ, it is well known that the unconfined compressive strength of the treated soil manufactured in-situ is usually much smaller than that manufactured in a laboratory. Figure 3.18 shows the relationship between the unconfined compressive strengths of treated soils manufactured in-situ, qu_f, and in the laboratory, qu_l. In the case of on land construction (Fig. 3.18(a)), the qu_f value may be as small as 1/2 - 1/5 of the laboratory strength, qu_l. In the case of marine construction (Fig. 3.18(b)), on the other hand, the qu_f value is of almost the same order of the laboratory strength, qu_l. The reason why the ratio of qu_f/qu_l is quite different in on land construction and marine construction is examined to be a relatively large amount of treated soil is manufactured with relatively good mixedness in marine construction (see Chapter 6).

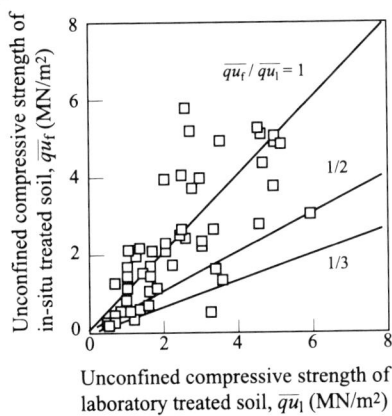

(a) treated soils of on land construction
(Public Works Research Center, 1999)

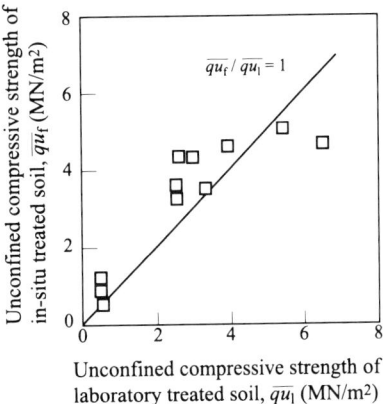

(b) treated soils of marine construction
(Noto et al., 1983)

Figure 3.18. Relationship between unconfined compressive strength of laboratory treated soil and of in-situ treated soil.

(3) *size effect on unconfined compressive strength*. In Japan unconfined compression tests are often conducted on a small specimen of 5 cm in diameter and 10 cm in height to estimate the compressive strength of a large scaled improved soil column. The Building Center of Japan (BCJ) conducted a series of compression tests on five different cement treated soils in order to investigate the size effect on the strength. In the tests, unconfined compression tests on in-situ columns of about 100 cm in diameter excavated from field and on small specimens with 5 cm in diameter sampled from the in-situ column were carried out. Figure 3.19 shows the test results, the averages of unconfined compressive strength on core samples and on full-scale samples are plotted on horizontal and vertical axes respectively. Although there is a scatter in the test data, it can be concluded that the unconfined compressive strength on the full-scaled column is about 70% of the average unconfined compressive strength on core specimen.

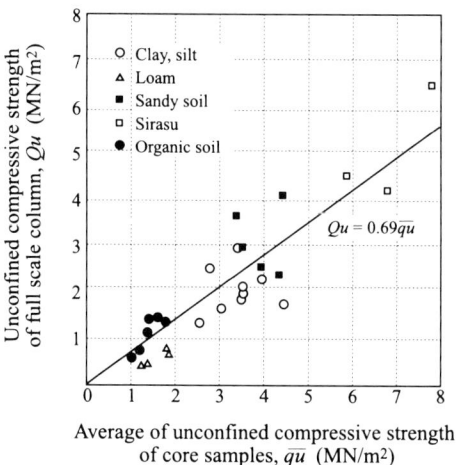

Figure 3.19. Scale effect on unconfined compressive strength (The Building Center of Japan, 1997).

References

Coastal Development Institute of Technology. 1999. *Deep mixing method technical manual for marine works*: 147 (in Japanese).

Hirade, T., M. Futaki, K. Nakano & K. Kobayashi 1995. The study on the ground improved with cement as the foundation ground for buildings, part 16. Unconfined compression test of large scale column & sampling core in several fields. *Proc. of the Annual Conference of Architectural Institute of Japan:* 861-862 (in Japanese).

Japan Cement Association. 1994. *Soil improvement manual using cement stabilizer*: 424 (in Japanese).

Kamata, H. & H. Akutsu. 1976. Deep mixing method from site experience. *Proc. of the Journal of Japanese Society of Soil Mechanics and Foundation Engineering, Tsuchi to Kiso*, 24 (12): 43-50 (in Japanese).

Kawasaki, T., A. Niina, S. Saitoh & R. Babasaki. 1978. Studies on Engineering Characteristics of cement-base stabilized soil. *Takenaka Technical Research Report*, 19: 144-165 (in Japanese).

Kawasaki, T., A. Niina, S. Saitoh, Y. Suzuki & Y. Honjo. 1981. Deep mixing method using cement hardening agent. *Proc. of 10th Internal Conference on Soil Mechanics and Foundation Engineering*, (3): 721-724

Niina, A., H. Saitoh, R. Babasaki, I. Tsutsumi & T. Kawasaki. 1977. Study on DMM using cement hardening agent (Part 1). *Proc. of 12th the Japan National Conference on Soil Mechanics and Foundation Engineering*: 1325-1328 (in Japanese).

Noto, S., N. Kuchida & M. Tereshi. 1983. Actual practice and problems on the deep mixing method. *Proc. of the Journal of Japanese Society of Soil Mechanics and Foundation Engineering, Tsuchi To Kiso*, 31(7):73-80 (in Japanese).

Public Works Research Center. 1999. *Deep mixing method design and execution manual for land works*: 47 (in Japanese).

Saitoh, S. 1985. Mechanical property of treated soil by the Deep Mixing Method, *Kisoko*, 13(2): 108-114 (in Japanese).

Sugiyama, K., T. Kitawaki & T. Morimoto. 1980. Soil Improvement Method of Marine Soft Soil by Cement Stabilizer. *Doboku Sekou*, 21(5): 65-74 (in Japanese).

Tatsuoka, F. & A. Kobayashi. 1983. Triaxial strength characteristics of cement treated soft clay. *Proc. of the 8th European Regional Conference on Soil Mechanics and Foundation Engineering*, 1:421-426.

Terashi, M., T. Okumura & T. Mitsumoto. 1977. Fundamental properties of lime treated soil (1st report). *Report of the Port and Harbour Research Institute*, 16(1): 3-28 (in Japanese).

Terashi, M., H. Tanaka, T. Mitsumoto, Y. Niidome & S. Honma. 1980. Fundamental Properties of lime and cement treated soils (2nd report). *Report of the Port and Harbour Research Institute*, 19(1): 33-62 (in Japanese).

Terashi, M., H. Tanaka, T. Mitsumoto, S. Honma & T. Ohhashi. 1983. Fundamental properties of lime and cement treated soils (3rd Report). *Report of the Port and Harbour Research Institute*, 22(1): 69-96 (in Japanese).

Terashi, M. & M. Kitazume. 1992. An investigation of the long-term strength of a lime treated marine clay. *Technical Note of Port and Harbour Research Institute*, 732: 1-15 (in Japanese).

The Building Center of Japan. 1997. Guideline on design and quality control of cement improved ground for architecture: 473 (in Japanese).

Yoshida, N., G. Kuno & H. Kataoka. 1992. Long-term strength on cement treated soil by the shallow mixing method. *Proc. of the 27th Japan National Conference on Soil Mechanics and Foundation Engineering*:2323-2326 (in Japanese).

CHAPTER 4

Applications

4.1 PATTERNS OF APPLICATIONS

(1) *improvement patterns*. In Japan, the Deep Mixing Method (DMM) has usually been applied to the improvement of soft clays and organic soils for various purposes and in various ground conditions (Terashi et. al., 1979, Terashi & Tanaka, 1981, Kawasaki et. al., 1981). Depending on the purposes and ground conditions, several methods have been conceived, as introduced in Figures 4.1 to 4.5. A suitable method should be chosen considering the purpose of improvement, stability calculation, construction cost, and site condition.

a) *block-type improvement*. In the block-type improvement, a huge improved soil mass is formed in the ground by overlapping every stabilized column (Fig. 4.1). This improvement can achieve the most stable improvement, but the cost is rather higher than the other improvements. This type of improvement is normally applied to heavy permanent structures such as the port and harbor structures.

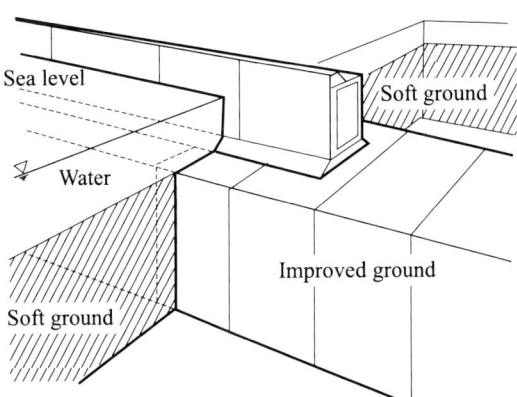

Figure 4.1. Block-type improvement.

b) *wall-type improvement*. In the wall-type improvement, the improved ground consists of long and short soil walls, where each wall is formed by overlapping every stabilized column (Fig. 4.2). The long wall has the function to bear the weight of superstructures and other external forces and transfers them to the deeper rigid ground layer. The short wall has the function to combine the long walls in order to increase the rigidity of the total improved soil mass. In recent times this improvement has been commonly applied in port and harbor constructions because of its reasonable cost.

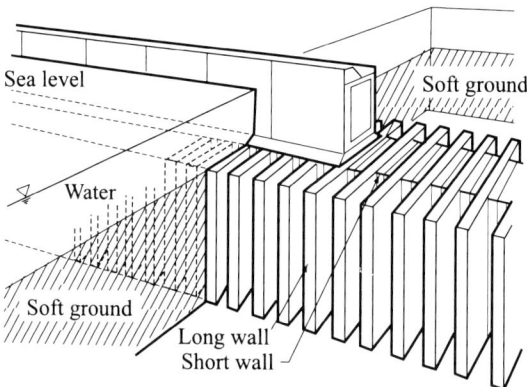

Figure 4.2. Wall -type improvement.

c) *lattice-type improvement (grid-type)*. The lattice-type improvement is recognized as an intermediate type between the block-type improvement and the wall-type improvement. The stabilized columns are constructed so that a lattice-shaped improved mass is formed in the ground (Fig. 4.3). This improvement has usually been applied to soil improvement under sea revetment, but recently it is also applied as a counter measure against ground liquefaction.

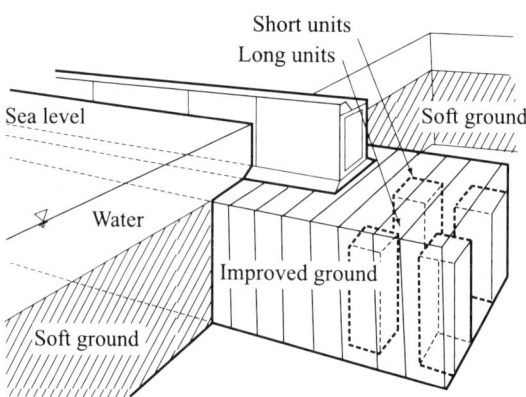

Figure 4.3. Lattice-type improvement.

d) *group column-type improvement*. In the group column-type improvement, many stabilized columns are constructed in rows with rectangular or triangular arrangements in the ground, where each column consists of several units of single soil columns linked together. In Japan, this improvement has been widely applied to relatively low embankments, foundations of light weight structures and temporary structures, in order to reduce settlement and/or to increase the stability (Fig. 4.4).

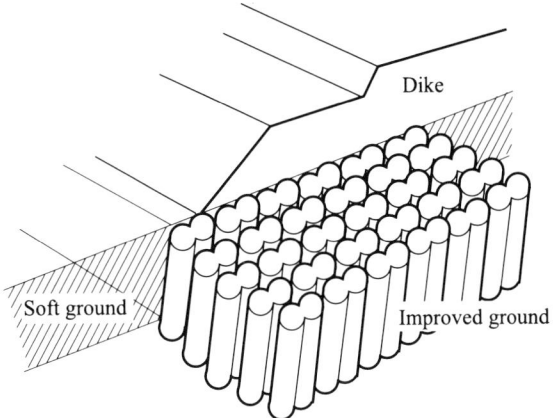

Figure 4.4. Group column-type improvement.

e) *columns in contact-type improvement*. This is a modified method of the group column improvement, where stabilized columns are in contact with the adjacent columns without overlapping. This improvement has larger bearing capacity against horizontal forces than the group column-type improvement.

Table 4.1 shows a comparison of the characteristics of the above mentioned improvements. It is concluded that the block-type improvement achieves the most stable improvement, but it is expensive. The wall-type improvement and the lattice-type improvement also achieve stable improvement, and are more economical, but both require high quality continuous overlapping work. Figure 4.5 shows the plan layout patterns of stabilized columns for all improvement methods, where a DM machine with double mixing shafts is adopted.

56 *The Deep Mixing Method – Principle, Design and Construction*

Table 4.1. Characteristics of improvement types.

	Stability	Cost	Installation	Design Consideration
Block Type	Large solid block resists external forces. Highly stable.	Volume of improvement is greater than other types. High cost.	Takes longer time because all columns are overlapped.	Design of size of block is in the same way as the gravity structures.
Wall Type	Where all improved walls are linked firmly, high stability is obtained.	Volume of improvement is smaller than block type. Lower cost.	Requires precise operation of overlapping of long and short units.	Requires consideration of unimproved soil between walls. Improvement size affected by internal stability.
Lattice Type (Grid Type)	Highly stable next to Block Type.	Cost range is between Block type and Wall type.	Installation sequences are complicated because lattice shape must be formed.	Requires design on three-dimensional internal forces.
Group Column Type	Where lateral forces are small, high stability is obtained.	Installation requires short period, and volume of improvement is small. Low cost.	Overlapping operation is not required.	Requires design on overall stability and on internal force of piles.
Columns in Contact Type	Where lateral forces are small, high stability is obtained. Stability increases by overlapping rows in direction of main external forces.	Cost is lower than Block type.	Precise operation is required to achieve firm contact of columns. Longer period is required if overlapping is required.	Requires design on overall stability and on internal force of piles.

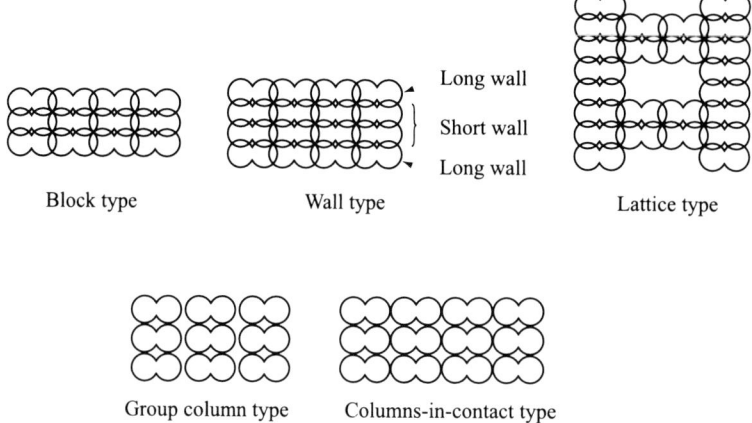

Figure 4.5. Arrangement of typical improvement pattern.

(2) *typical applications of DMM*. Figure 4.6 shows typical applications of DMM in Japan. The purpose of these applications is varied, such as reducing settlement, increasing bearing capacity of the ground, increasing stability and ground liquefaction strength, reducing active earth pressure, cutting off ground water, and increasing pile's bearing capacity against lateral forces, etc. Figures 4.7 to 4.10 show typical applications of DMM.

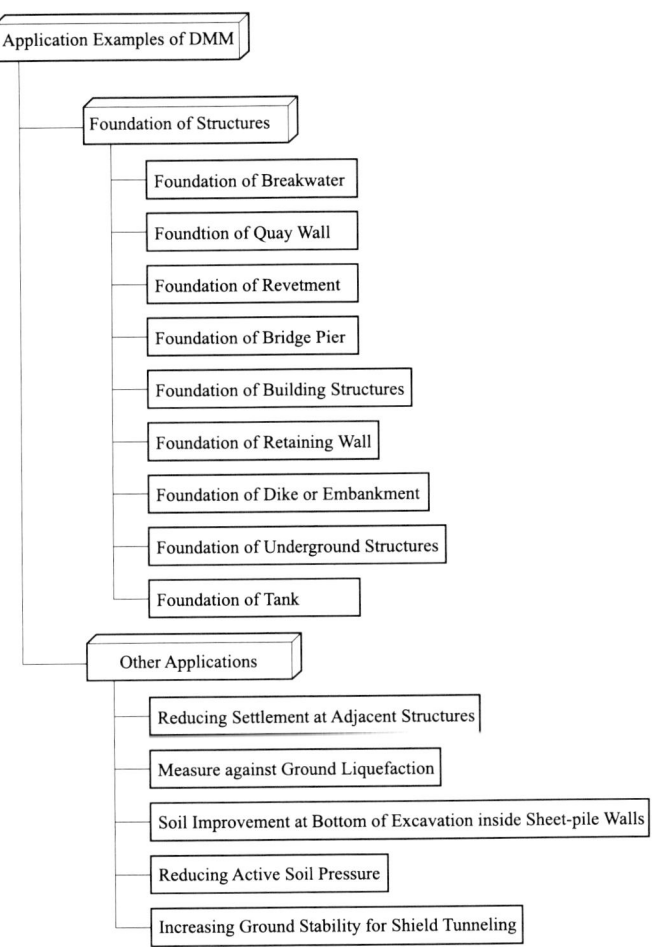

Figure 4.6. Various applications of DMM.

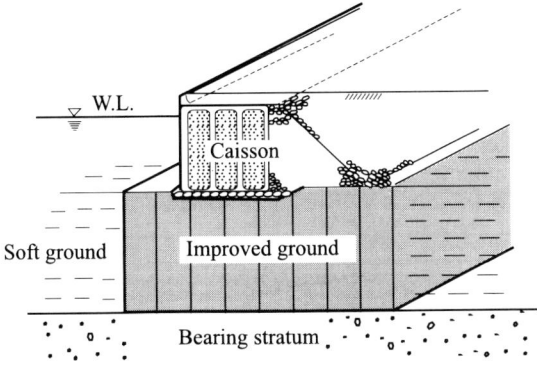

Figure 4.7. Ground improvement for caisson-type sea wall.

58 *The Deep Mixing Method – Principle, Design and Construction*

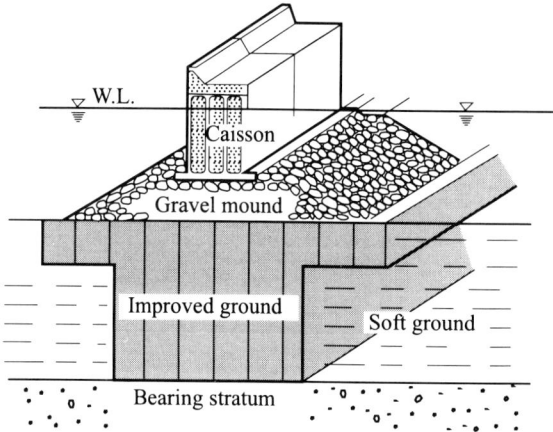

Figure 4.8. Ground improvement for breakwater.

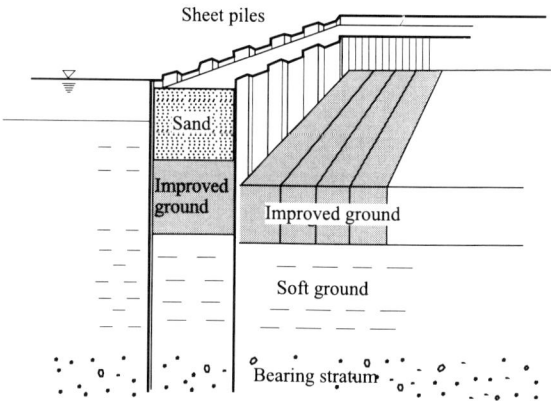

Figure 4.9. Soil improvement at bottom of excavation inside sheet pile walls.

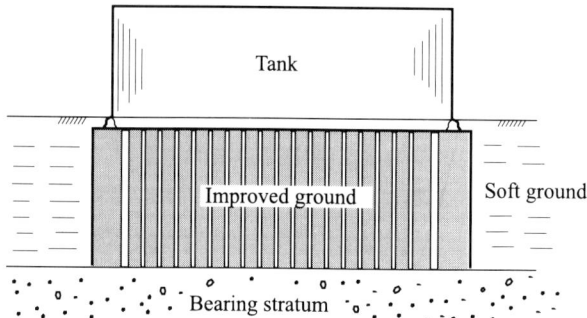

Figure 4.10. Ground improvement for tank support.

4.2 APPLICATIONS IN JAPAN

In Japan, the accumulative volume of treated soil using wet-type DMM from 1977 to 1998 reached 38 million cubic meters. Figure 4.11 shows a comparison of the treated soil volumes between on land application and marine application. For on land applications, the method has mainly been applied to improve slope stability, to prevent building subsidence and to improve the bearing capacity of foundations. In approximately 50% of marine applications, it has been applied to improve the foundations of revetments.

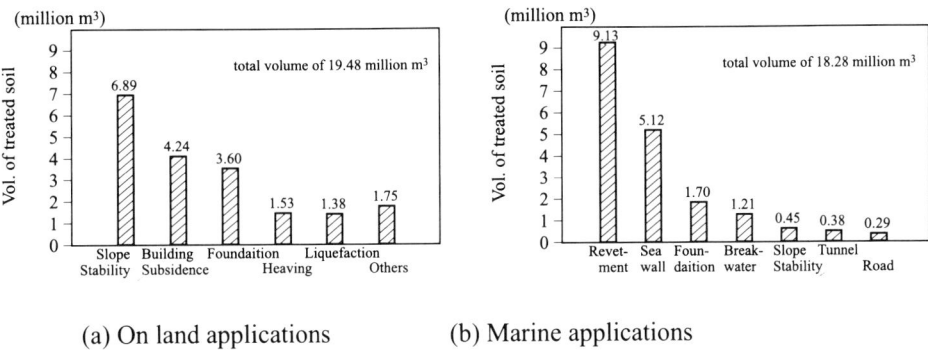

(a) On land applications   (b) Marine applications

Figure 4.11. Volume of treated soil for on land and marine applications.

Among a lot of previous applications of DMM in Japan, four examples are selected and briefly introduced in this section: Trans-Tokyo Bay Highway Project, Kansai International Airport Project, Yodo River Dike Project and Breakwater Construction Project at Fukushima Port (Fig. 4.12).

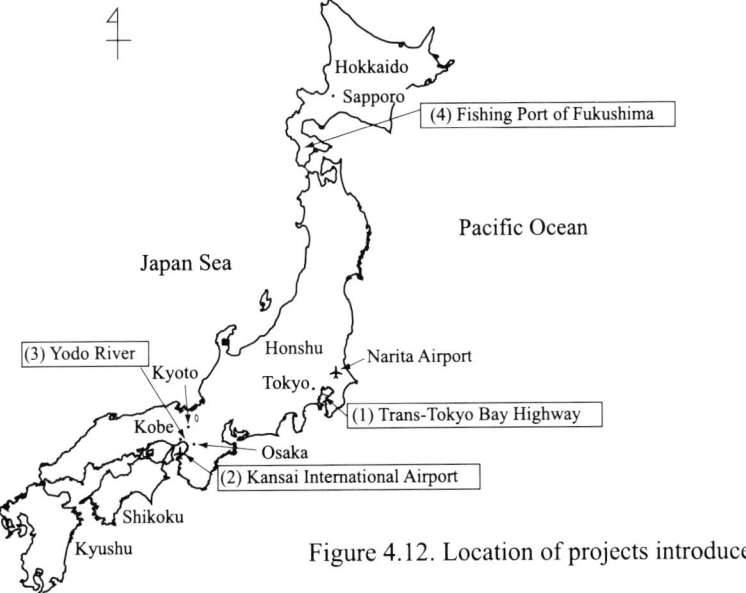

Figure 4.12. Location of projects introduced.

(1) *Trans-Tokyo Bay Highway Project*. The Trans-Tokyo Bay Highway Project was planned to improve the heavy traffic condition of Metropolitan Tokyo by creating a by-pass between Kanagawa and Chiba prefectures. In this project, a new 15.1 km-long highway as well as two man-made islands was constructed in Tokyo Bay, as shown in Figure 4.13 (Uchida et al., 1993, Tatsuoka et al., 1997, TTBHC, 1998). The Deep Mixing Method was applied at three construction sites; Ukishima Access, Kawasaki island and Kisarazu island. The project was successfully completed in 1997.

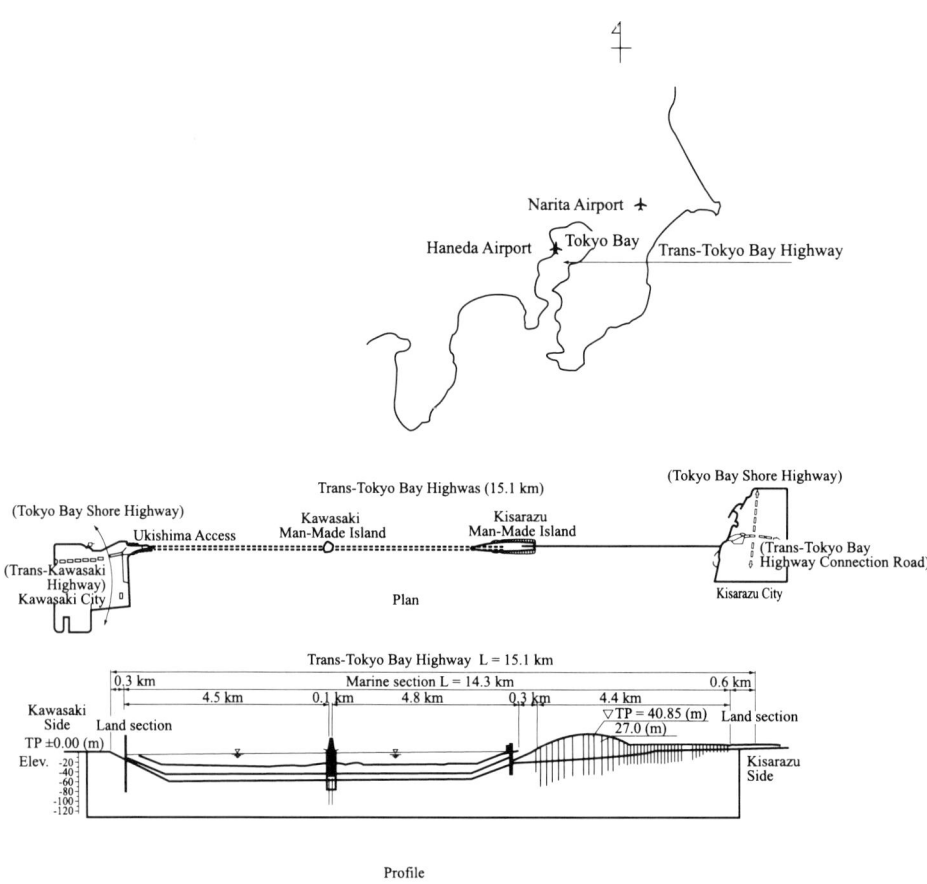

Figure 4.13. General plan/ profile of Trans-Tokyo Bay Highway.

Figure 4.14 shows a plan view of Ukisima Access and its cross section. Ukishima Access was constructed using an underwater embankment. The figure shows the ramp section entering the sea from the front face of the ventilation tower. There is a sloped embankment measuring 100 m wide and roughly 700 m long, in which the shield tunnel is contained. At the site location, the upper 30 m of the sea bed foundation is an extremely soft alluvial clay. To reduce the consolidation settlement, and to ensure the necessary stability to safeguard the tunnel, the ground in the center part was improved by DMM with cement slurry. The outside area was also improved by the sand compaction pile method (SCP) to increase the lateral resistance of the jacket wall.

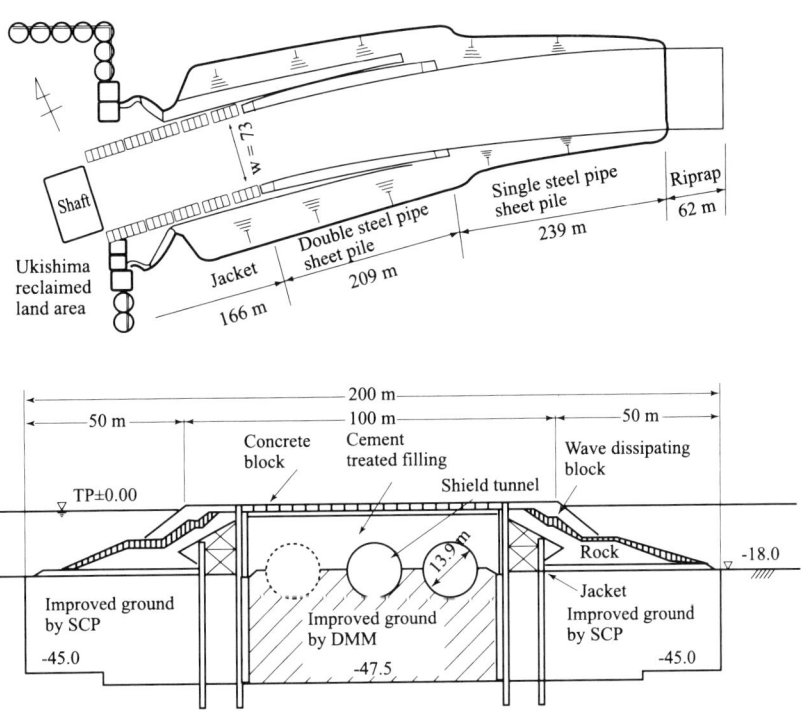

Figure 4.14. Plan view and cross section of Ukishima Access.

Figure 4.15 shows the data of natural water content and unit weight at Ukishima site. The clay layer from -20 m (seabed) to about -40 m in depth has a water content of about 120%. The unconfined compressive strength of the natural ground increases with the equation, $qu = 4.3 \times z - 86$ (kN/m^2), where z is the depth of the ground. It was planned to improve the ground up to −45 m for the construction of shield tunnels. A total volume of 1.3 million m^3 was improved at Ukishima site by using the wet type Deep Mixing Method. The strength of improved ground was designed at an average strength of 1 MN/m^2 for smooth excavation and for the stability of the front cutting face of the shield tunnel. Based on the laboratory mixing test results, a cement content of 70 kg/m^3 and water/cement ratio w/c of 100% were determined to obtain the design strength. To ensure good mixing conditions, a field execution test was also performed at Ukishima Access construction site.

Figure 4.15 also shows the data of 28 days' unconfined compressive strength of the improved ground at Ukishima Access. Although field data are scattered over a wide range, it can be seen that the average strength of the improved ground satisfies the design value. Triaxial compression tests were also conducted on the stabilized soil and their results were also plotted in the figure. The strength obtained by the triaxial compression tests was larger than the strength by unconfined compression tests, and the data were less scattered compared to the unconfined compressive strength.

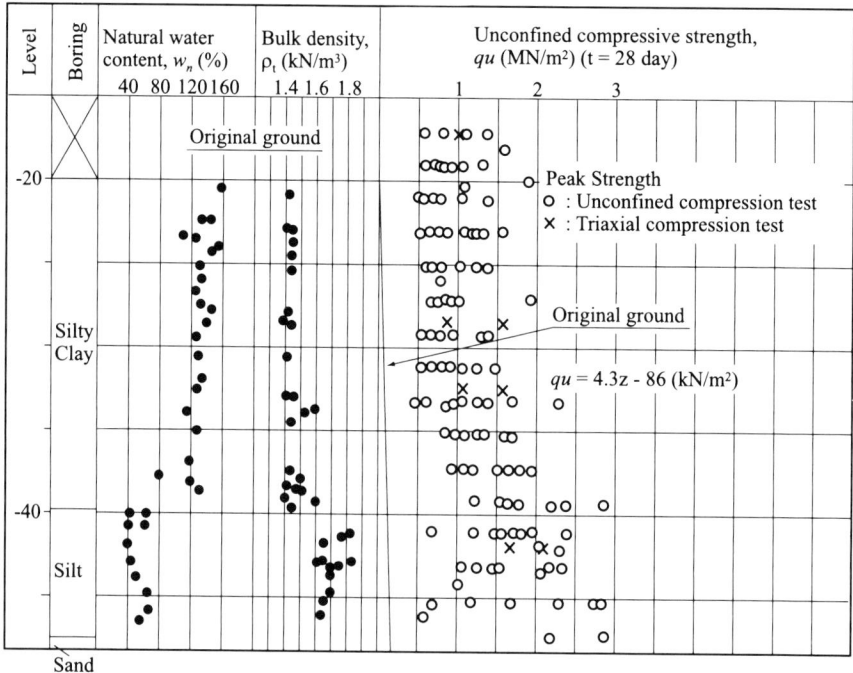

Figure 4.15. Result of investigation from actual work.

(2) Kansai International Airport Construction Project. Kansai International Airport was opened as Japan's first 24-hour airport in September 1994. The airport construction involved the reclamation of a large scaled man-made island in Osaka Bay. Although the airport in the first phase is 5.1 km² and has only one runway now, the total area of the final phase will be 12 km² to cover the growing air traffic demand. It is located 5 km off the mainland to avoid noise problems.

Because the foundation condition is very weak, the consolidation settlement is still ongoing and at about 40 cm per year even after the airport opened. The reclaimed island was constructed in seawater with an average depth of 18 m, and under the island there is a 20 m thick soft alluvial clay deposit. Below the alluvial clay layer, thick diluvial layers are deposited in alternating clay, sand and gravel layers, with a depth of several hundred meters. Because the diluvial layers are too deep and too thick, it is impossible to prevent their consolidation settlement.

The alluvial clay layer has mainly been stabilized by the consolidation technique in the island part. The revetment or seawall surrounding the island consists of rubble mounded structures or steel cell structures laid on the improved clay soil. As ground improvement techniques, the sand drain and sand compaction pile methods were predominantly used. About 4 to 5 m consolidation settlement took place during the construction period.

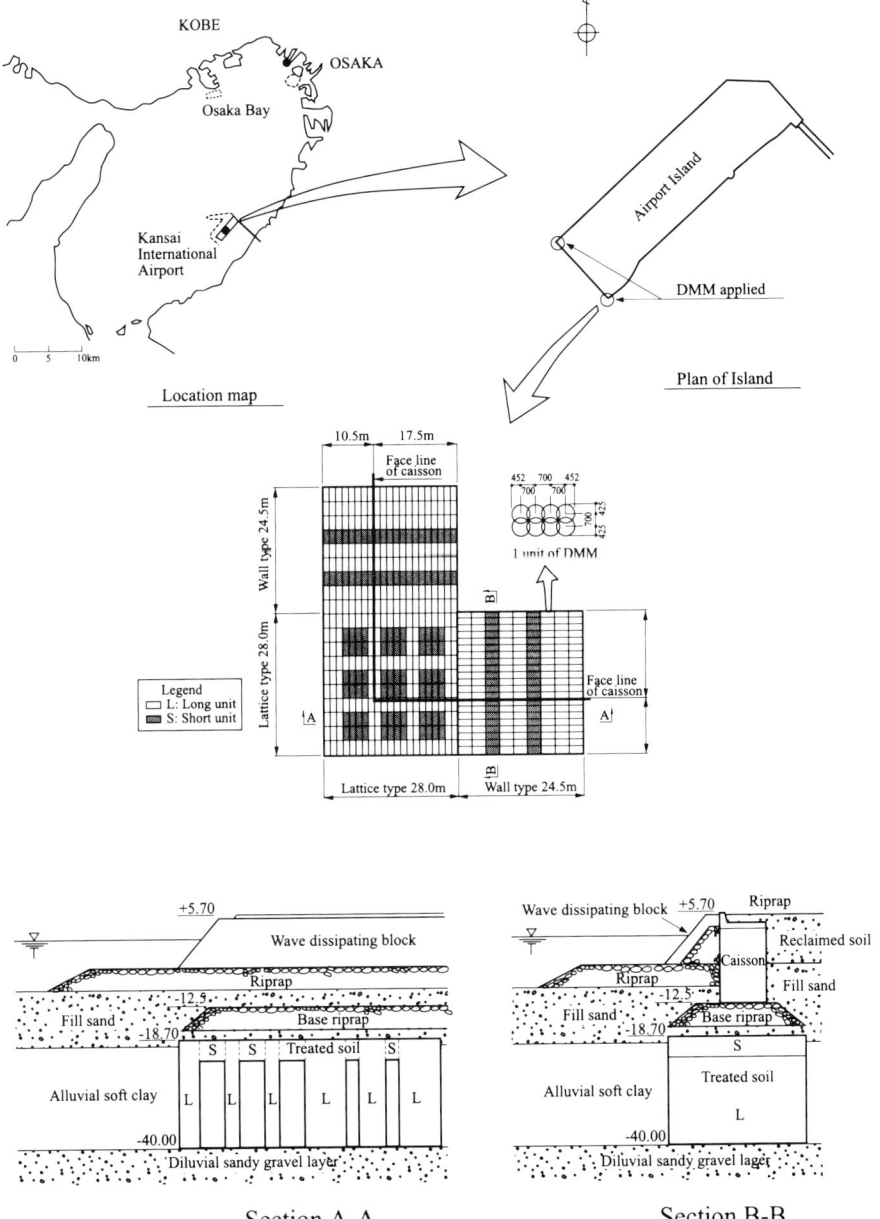

Figure 4.16. Ground improvement for Kansai International Airport.

At the corner of the seawall, DMM with wall-type improvement was applied to the alluvial clay layers in order to reduce the construction period. In this method of improvement, 3 to 9 m-thick long walls were constructed at 10 m intervals, and were connected with short walls. In order to confirm the appropriate design method, a full-scale test was conducted on reclaimed land in Osaka, in which the wall-type improved ground was subjected to backfill to cause failure.

At the corner of the seawall, the structure must bear the external forces from various directions. It is necessary therefore to connect all stabilized columns, avoiding discontinuous joints. In soil improvement practice, however, it is difficult to form a large continuous block because a stabilized column will be installed in an overlapping fashion before hardening of the prior installed stabilized columns, where the improvement plant moves row by row. To solve this problem, newly developed slow-hardening cement was used and the sequence of improvement work was carefully studied. Even in such severe conditions, the DMM execution was successfully completed and additional settlement was prevented.

(3) *Yodo River Dike Project.* Yodo River flows from Lake Biwa to Osaka Bay through Osaka City. Due to the Hyogoken-Nambu Earthquake in January 1995, the river dike was damaged for the length of 1.8 km because of slope failure due to ground liquefaction (Kamon,1996). A representative cross section of the damaged dike is shown in Figure 4.17. The top portion of the river dike sank down about 3 m. The damaged dike had to be restored very quickly because there was a risk of flooding during the rainy season which commences in June. Because there were many residential houses in the neighborhood along the river dike, it was necessary to avoid noise and vibratory problems during the construction. Therefore, the DMM method was mainly applied there because it has less noise and vibration. Grid type improvement was applied to increase the stability of the dike, as shown in Figure 4.18. In the construction period, more than 50 DMM machines were simultaneously put into operation for rapid restoration.

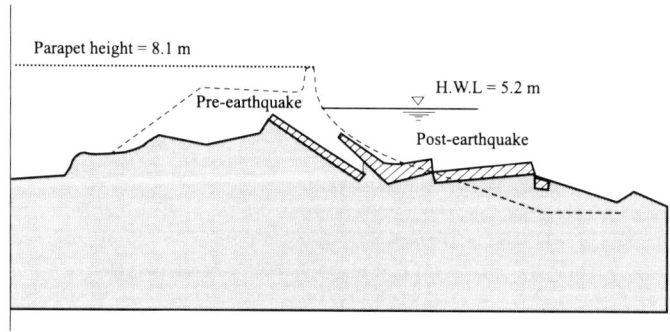

Figure 4.17. Earthquake-damaged dikes along Yodo River.

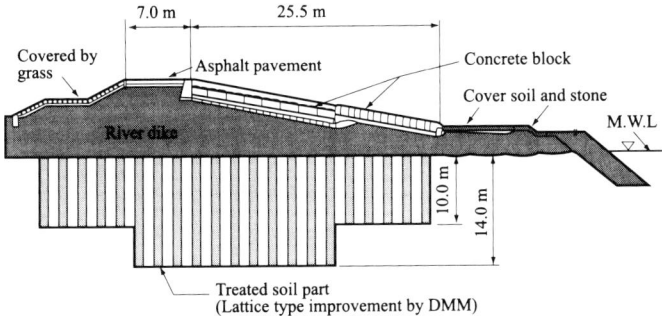

Figure 4.18. Cross section of restored river dike by DMM.

(4) *Breakwater construction project at Fukushima Port*. A breakwater with 152 m in length was constructed at Fukushima Port, Hokkaido in the winter of 1995. In this project, the soft clay layer was improved by DMM with cement slurry (Tokoro & Ogura,1997). Figure 4.20 shows a typical cross section of the breakwater. This was the first application of DMM under very cold climatic conditions in Japan. During the construction, the lowest temperature was -12 °C, and snowfall was 135 cm in a month. All the plant and equipment were lubricated with non-freeze oil and grease and the pipes on the decks of work vessels were heated. In order to estimate the influence on the workability of the slurry of the low temperature, preliminary tests were performed under similar weather conditions. Considering the test

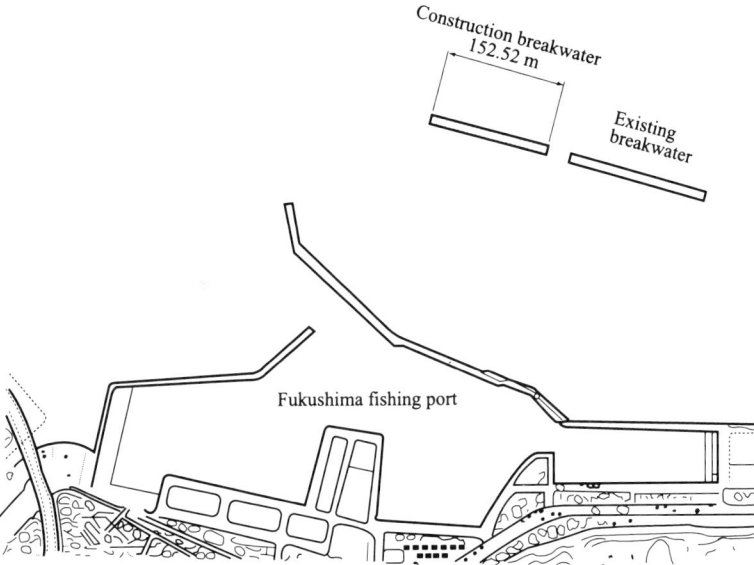

Figure 4.19. Planning area of DMM.

results, the pouring time of the slurry was controlled to be within 60 minutes after mixing. The Global Positioning System (GPS) was applied for positioning the DM barge because optical survey equipment could not be used during heavy snowy weather. Although the improvement work was sometimes interrupted by heavy storms, the construction work was completed successfully with relatively slow progress.

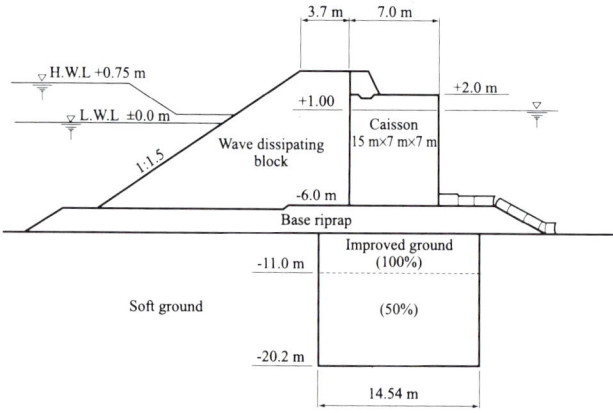

Figure 4.20. Cross section of breakwater.

Figure 4.21. Photo of DMM barge.

References

Kamon, M. 1996. Effect of grouting and DMM on big construction projects in Japan and the 1995 Hyogoken-Nammbu earthquake. *Proc. of the Second International Conference on Ground Improvement Geosystems,* 2: 807-823.

Kawasaki, T., A. Niina, S. Saitoh, Y. Suzuki & Y. Honjo. 1981. Deep mixing method using cement hardening agent. *Proc. of the 10th International Conference on Soil Mechanics and Foundation Engineering*, 3: 721-724.

Tatsuoka, F., U. Uchida, K. Imai, T. Ouchi & Y. Kohata. 1997. Properties of cement treated soils in Trans-Tokyo Bay Highway Project. *Journal of Ground Improvement*, (1)1:37-57.

Terashi, M., H. Tanaka & T. Okumura. 1979. Engineering properties of lime treated marine soils and the DM method. *Proc. of the 6th Asian Regional Conference on Soil Mechanics and Foundation Engineering*, 1: 191-194.

Terashi, M. & H. Tanaka. 1981. Ground improved by the deep mixing method. *Proc. of the 10th International Conference on Soil Mechanics and Foundation Engineering*, 3: 777-780.

Tokoro, E. & T. Ogura. 1997. Case histories of CDM in the offshore in severe winter. *Proc. of the Engineering Report on The memorial 20 years*. The Cement Deep Mixing Method Association: 20-23 (in Japanese).

TTBHC. 1998. Trans-Tokyo Bay Highway Project: 177 - 223 (in Japanese).

Uchida, K., Y. Shioji & Y. Kawase. 1993. Cement treated soil in the Trans-Tokyo Bay Highway project. *Proc. of the 13th International Conference on Soil Mechanics and Foundation Engineering*: 1179-1182.

CHAPTER 5

Design of Improved Ground by DMM

5.1 INTRODUCTION

A column of treated soil is constructed by the operation of a DM machine, and a treated soil mass with any desired shape can be formed in the ground by overlapping these columns. Figure 5.1 shows typical improvement patterns of a treated soil mass (Terashi & Tanaka, 1981). To improve the foundation ground for a permanent and/or important structure, block-, wall- or lattice- (grid-) type improvements have frequently been applied to port and harbor facilities in Japan. The group column-type treatment pattern has usually been applied to light weight or temporary structures or embankment construction in order to reduce vertical and horizontal displacement, because each stabilized column is anticipated to fail progressively by bending as a result of the low tensile and bending strength of the treated soil (Karastanev et al. 1997, Kitazume et al., 2000). Consequently, this improvement pattern has rarely been applied to the construction of harbor and port facilities that will be subjected to large horizontal loads.

The DM method is now widely applied in Japan to port facilities, embankments, excavations, and so on. In this chapter the current design procedures for block-, wall-, and group column-type improvements are briefly described. The design procedures introduced in the following sections are still being modified in accordance with the new findings from many research projects and accumulated experience, which include the design load conditions for external and internal stability analyses, the static and dynamic behaviors of the improved ground, and a great deal of know-how in design and practice. Therefore a specific practical design for each site should be produced not only according to the procedure, but also in accordance with ongoing research activities and accumulated know-how.

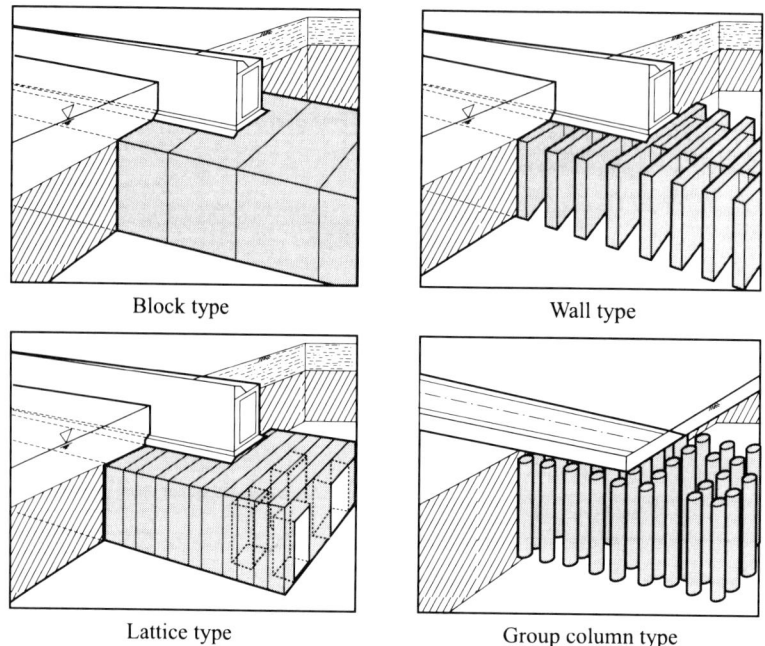

Figure 5.1. Typical improvement patterns of treated soil mass.

5.2 DESIGN PROCEDURE FOR BLOCK-TYPE AND WALL-TYPE IMPROVEMENTS

(1) *basic concept*. Japanese marine clays treated by DMM are well known to have high strength of the order of 1 MN/m^2, small strain at failure of the order of 0.1%, relatively low tensile strength, low permeability, and improved consolidation characteristics (Terashi et. al., 1979 & 1980, Kawasaki et al., 1981). The strength of in-situ treated soils deviates markedly from the average value, as shown in Figure 5.2 (Nakamura, 1977), even if the manufacturing has been performed with the greatest possible care. Furthermore, when the stabilized soil columns are overlapped to make a continuous treated soil mass, the boundary surfaces between adjacent columns (a sort of construction joint) may become weak points in a structure. Therefore, sufficiently high safety factors are applied to the strength of in-situ treated soil; this in turn results in extraordinary difference between the engineering characteristics of the treated soil and untreated surrounding soft soil. A typical example of the stress-strain curve of treated soil is shown in Figure 5.3 together with that of untreated soil (Sugiyama et al., 1980). It is therefore inappropriate to assume that DMM-improved ground is similar to ground treated with other soil improvement techniques such as the vertical drain method or the sand compaction pile method. Hence, when block type improvement is applied to the improvement of port facility construction, the treated soil is not considered to be a part of the ground, but rather to be a rigid structural member buried in the

ground to transfer the external loads to a reliable stratum (Ministry of Transport, 1999).

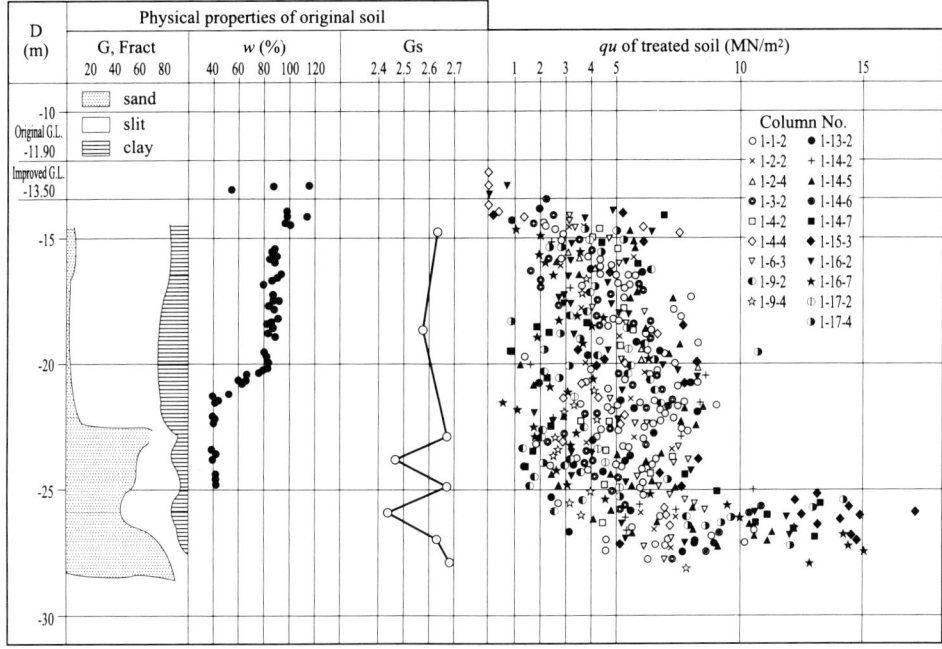

Cement content α = 160 kg/m³
G. Fract. denotes percentage of each grain size fraction

Figure 5.2. Unconfined compressive strength of in-situ treated soil at Yokohama Port (Nakamura et al., 1980).

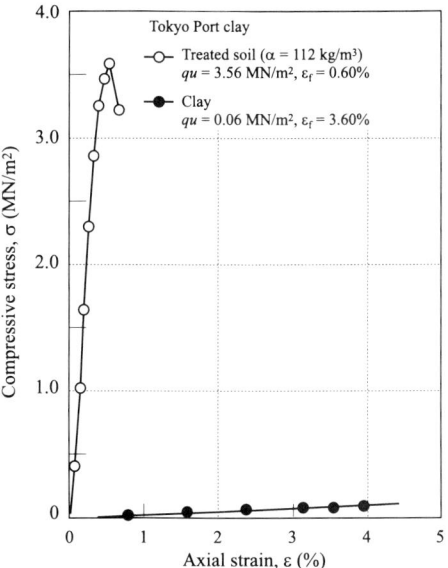

Figure 5.3. Stress-strain curve of treated soil (Sugiyama et al., 1980).

72 *The Deep Mixing Method – Principle, Design and Construction*

(2) *design procedure*. The current design procedure for the block- and wall-type improvements of port facilities which was established by the Ministry of Transport and is widely applied to constructions throughout Japan, is shown in Figure 5.4 (Ministry of Transport, 1999). In the wall type improvement composed of long and short walls as shown in Figure 5.4, the basic design concept can be assumed to be similar to the block type improvement, because the long walls function to transmit external force generated by the superstructure to the supporting ground and the short walls function to join the long walls together. The design concept is, for the sake of simplicity, derived by analogy with the design procedure for a gravity-type structure such as a concrete retaining structure.

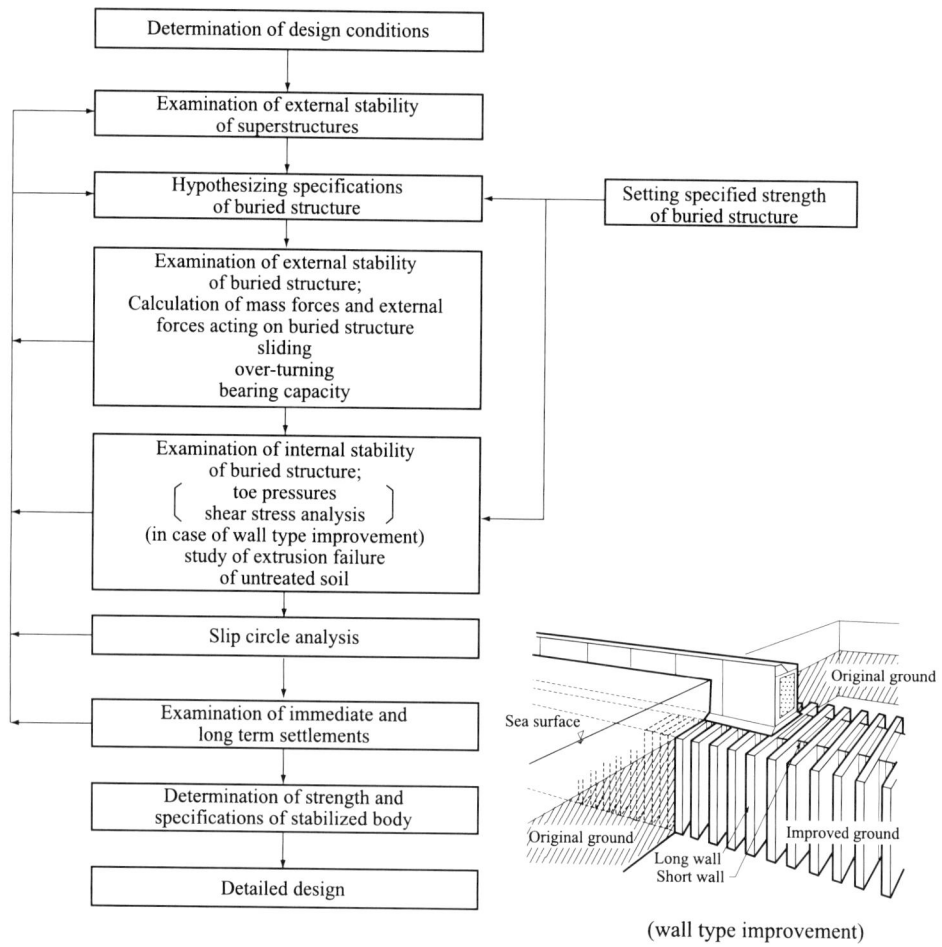

Figure 5.4. Flow of the current design procedure (Ministry of Transport, 1999).

The first step in the procedure is stability analysis of the superstructure to ensure the superstructure and DMM-improved ground can behave as a whole. The second step is an "external stability analysis" of improved ground manufactured in-situ by DMM in which sliding failure, overturning failure and bearing capacity of the DMM ground are evaluated. The third step is an "internal stability analysis" of a buried rigid structure, in which the induced stresses within the DMM due to the external forces are calculated and confirmed to be less than the allowable value. The wall-type improved ground is also checked for extrusion failure, where untreated soil within the long walls might be pushed out. The fourth step is an examination of the displacement analysis of DMM-improved ground.

(a) For the stability analysis of the superstructure at the first step of the design procedure, the improved ground that is not yet determined is assumed to have adequate bearing capacity to support the superstructure. The sliding and overturning failures of the superstructure are calculated at this step. In the seismic design of the superstructure, the seismic intensity analysis is applied in Japan; the dynamic cyclic loads are converted to static load by multiplying the unit weight of the structure by the seismic coefficient. The design seismic coefficient of the superstructure can be obtained by Equation (5.1). As it is found that the DMM-improved ground has better seismic characteristics than the original ground, the modification factor for stiff ground can be used in Eq. (5.1).

Design seismic coefficient = Seismic intensity by zone ×
  Modification factor for ground ×
  Modification factor for importance · · · · · · · · · · · · · · · · · Eq. (5.1)

(b) In the "external stability analysis" of a rigid buried structure, three failure modes are examined for the assumed cross section of the treated soil mass: sliding, overturning and bearing capacity failures. Design loads considered in the external stability analysis are shown schematically in Figure 5.5. These include active and passive earth pressures, other external forces that are applied to the boundary of the treated soil structure, the mass forces generated by gravity, and the inertia generated by an earthquake.

A series of centrifuge model tests reveals that a certain degree of cohesion is mobilized on the active and passive side surfaces of the improved ground when it displaces (Kitazume, 1994). Cohesion may be generated due to the consolidation settlement of the untreated soil by backfill on it. The cohesion mobilized in the downward direction on the active side surface of the improved ground functions to increase the stability of the improved ground. The cohesion is incorporated in the current design to examine the stability of the improved ground.

In the stability analysis of the wall-type improved ground, it is sometimes necessary to set the external forces separately for untreated soil and treated soil. In general, it can be assumed that the active and passive earth pressures act uniformly on the long wall and on the untreated soil, and the self weight of the superstructure and external forces acting on the superstructure and improved ground are concentrated on the long wall.

74 *The Deep Mixing Method – Principle, Design and Construction*

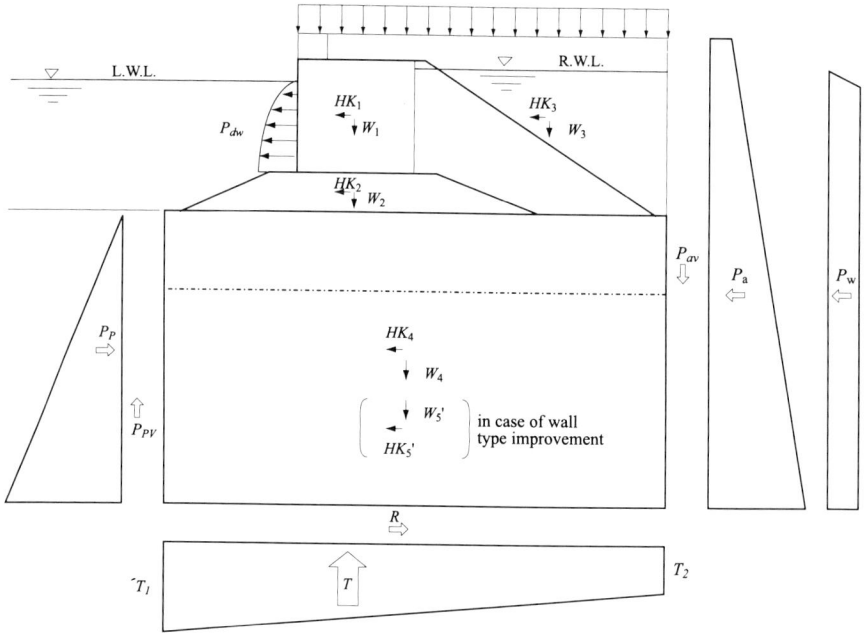

Figure 5.5. Schematic diagram of design loads (Ministry of Transport, 1989).

The sliding and overturning stabilities are calculated by the equilibrium of horizontal and moment forces, and the safety factors for these failures are calculated by Equations (5.2) and (5.3) respectively.

$$Fss = \frac{Pp + R}{Pa + Pw + \Sigma Hki} \quad \cdots\cdots\cdots\cdots\cdots\cdots\cdots\cdots\cdots\cdots\cdots \text{Eq.(5.2)}$$

$$Fso = \frac{Pp \times Yp + Pav \times Xav + \Sigma Wi \times Xi}{Pp \times Ya + Pw \times Yw + \Sigma Hki \times Yi} \quad \cdots\cdots\cdots\cdots\cdots \text{Eq. (5.3)}$$

where
- *Fso*: safety factor against overturning failure
- *Pa*: active earth pressure
- *Pp*: passive earth pressure
- *Pav*: vertical cohesion force of untreated soil acting on the active side surface
- *Ppv*: vertical cohesion force of untreated soil acting on the passive side surface
- *Pw*: residual water pressure
- *R* : shear strength acting on the bottom of the improved ground
- *ΣHki*: sum of inertial forces generated by an earthquake
- *ΣWi*: total of the masses

X: horizontal distance from the vertical force to the front edge of the improved ground
Y: vertical distance from the horizontal force to the bottom of the improved ground

In the bearing capacity analysis, the induced pressure distribution on the bottom of the improved ground is calculated based on the simple force and moment equilibrium of external forces, and is confirmed to be lower than the allowable bearing capacity. The induced pressure distribution is calculated by Equation 5.4 (see Fig. 5.5). The allowable bearing capacity is greatly dependent upon the soil condition overlaid by the improved ground. But in the case of sandy ground, the allowable bearing capacity in the current design is usually set up to be about 600 kN/m² and 900 kN/m² for static and dynamic conditions respectively.

if $e \leq B/6$

$$T_1 = \frac{W' \times (1 + 6 \times e/B)}{B}$$

$$T_2 = \frac{W' \times (1 - 6 \times e/B)}{B}$$

if $e \geq B/6$

$$T_1 = \frac{2 \times W'}{3 \times B}$$

.... Eq. (5.4)

where
B : width of the improved ground
e : eccentricity
T_1 : reaction pressure at front edge of improved ground
T_2 : reaction pressure at rear edge of improved ground
W' : sum of vertical forces
X : position of force intensity

The design loads acting on the improved ground for the bearing capacity analysis were originally considered to be the ultimate active and passive earth pressures, the same as in the "external stability analysis". However, in a case where the improved ground is sufficiently stable with some margin of the safety factor, it is easily understood that the earth pressure on the passive side of the improved ground and the shear strength on the bottom are not fully mobilized to the ultimate value. Recent research efforts have revealed this phenomenon experimentally and analytically (Kitazume, 1994). According to his investigation, the design load on the passive side for the bearing capacity analysis should be determined by considering the force equilibrium of loads acting on the active side and modified shear force on the bottom.

Since the magnitude and distribution of the earth pressures up to failure are still not well determined, detailed analysis such as FEM analysis should be conducted to achieve more economical and more accurate design.

The bearing capacity of a row of treated soil walls in the wall-type improved ground is a problem of mutual interference of the bearing capacities of deep rectangular foundations. The increase of the bearing capacity of treated soil walls caused by the interference of adjacent walls has been demonstrated in a series of centrifuge model tests and the simple design shown in Figure 5.6 has been proposed by Terashi & Kitazume (1987).

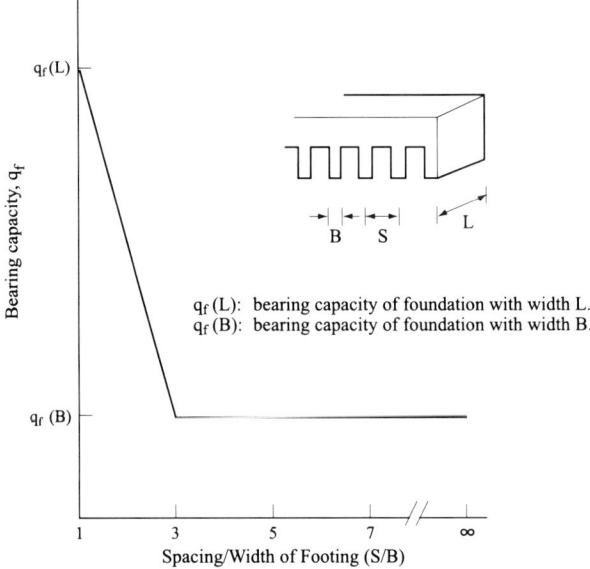

Figure 5.6. Simple design of bearing capacity of wall type improvement (Terashi & Kitazume, 1987).

(c) In the "internal stability analysis" of improved ground, the induced stresses in the improved ground are calculated based on elastic theory. The shape and size of the improved ground are determined so that the induced stresses become lower than the allowable strengths of the treated soil. At the moment, the allowable strengths of treated soil are defined by Equations (5.5) to (5.8).

$$\sigma_{ca} = \alpha \cdot \beta \cdot \gamma \cdot \overline{qu_f} / Fs \quad \cdots \quad \text{Eq. (5.5)}$$
$$\sigma_{ca} = \alpha \cdot \beta \cdot \gamma \cdot \lambda \cdot \overline{qu_f} / Fs \quad \cdots \quad \text{Eq. (5.6)}$$
$$\tau_a = \sigma_{ca} / 2 \quad \cdots \quad \text{Eq. (5.7)}$$
$$\sigma_{ta} = 0.15 \cdot \sigma_{ca} \leqq 200 kN/m^2 \quad \cdots \quad \text{Eq. (5.8)}$$

where
  τ_a: allowable shear strength
  σ_{ca}: allowable compressive strength

σ_{ta} : allowable tensile strength
$\overline{qu_f}$: average unconfined compressive strength of in-situ treated soil
$\overline{qu_l}$: average unconfined compressive strength of treated soil manufactured in a laboratory
α : coefficient of effective width of treated soil column (0.7 – 0.9)
β : reliability coefficient of overlapping (smaller than unity)
γ : correction factor for scattered strength (0.5 – 0.6)
λ : ratio of qu_f/qu_l (usually 0.5 to 1 according to past experience)
Fs: safety factor
    3.0 in ordinary condition
    2.0 in earthquake condition

a) *safety factor Fs of the materials*. As all the allowable strengths are based on the unconfined compressive strength, *qu*, in which no effect of creep and cyclic loading are incorporated. In the practical design procedure, safety factors of 3 and 2 for static and dynamic conditions respectively are set up to incorporate these effects, and also to incorporate the importance of the structure, the type of loads, the design method, and the reliability of the materials.

b) *coefficient of effective width of treated soil column*, α. Improved ground manufactured by serial overlapping is in general composed of stabilized columns and untreated soil within the columns, as shown in Figure 5.7. The coefficient of the effective width of treated soil column α is the coefficient used to compensate for this untreated part.

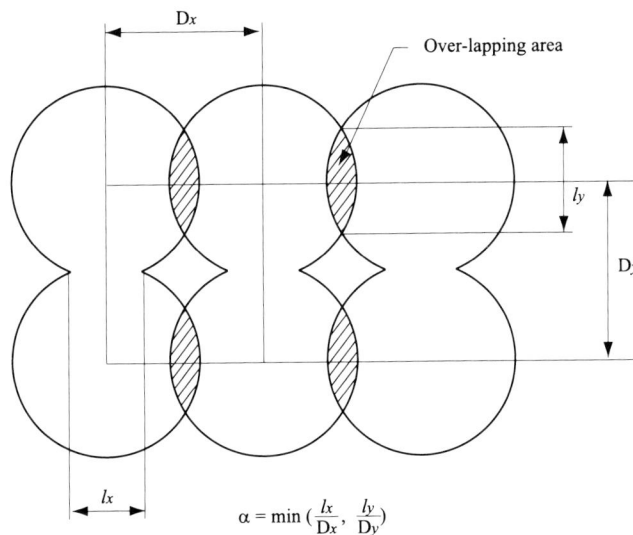

Figure 5.7. Effective width formed by improvement machine.

c) *reliability coefficient of over lapping, β*. In the overlapping execution, a stabilized column during hardening is partially scraped by the following column (Fig. 5.8). The strength in the overlapped area is anticipated to be lower than that of other parts of the column. The reliability coefficient of overlapping is the ratio of the strength of overlapped and non-overlapped areas. Its magnitude is influenced by various factors, such as the time interval until overlapping, the execution capacity of the DM machine, and the type of stabilizing agent discharged, but is usually set as 0.8 to 0.9.

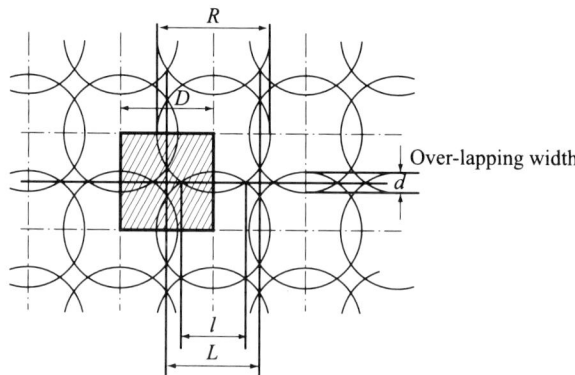

Figure 5.8. Connecting surfaces.

d) *correction factor for scattered strength, γ*. It is well known that unconfined compressive strength on in-situ treated soil has a large scatter. The correction factor for scattered strength is a coefficient used to account for this scattering.

c) *ratio of $\overline{qu_f}/\overline{qu_l}$, λ*. The accumulated data clearly shows that the average unconfined compressive strength of in-situ treated soil, qu_f has a clear relation to that of laboratory treated soil qu_l, as shown in Equation (5.9). The value of λ can be set at 1 in the design, when the improved ground is manufactured for a port facility using either ordinary Portland cement or Portland blast furnace slag cement (Fig. 5.9).

$$\overline{qu_f} = \lambda \, \overline{qu_l} \quad \cdots \cdots \cdots \cdots \cdots \cdots \cdots \cdots \cdots \cdots \cdots \cdots \quad \text{Eq. (5.9)}$$

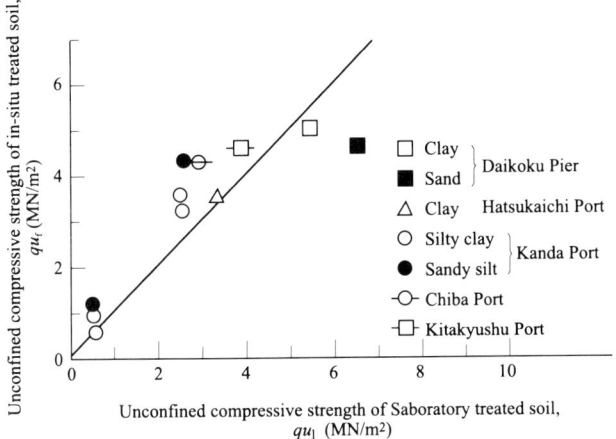

Figure 5.9. Relationship between unconfined compressive strength of in-situ treated soil, qu_f and laboratory treated soil, qu_l (Noto et al., 1983).

Since the coefficients described above are not independent but are actually closely related, it is difficult to determine their magnitude individually. According to previous case histories, the ratio of the allowable compressive stress to the unconfined compressive strength in laboratory treated soil is often between 1/6 and 1/10.

The loads applied to the internal stability analysis were generally assumed to be similar to those for the external stability analysis, as already shown in Figure 5.5. Yet Terashi et al. (1989) proposed, based on their centrifuge tests, that the earth pressures acting on the improved ground should be close to the pressures at rest in the internal stability analysis as long as the safety factor for external stability is relatively large. The design loads should therefore be carefully determined by considering the margin of the safety factor for external stability.

(d) *extrusion failure*. For wall-type improvements, the extrusion failure must also be examined. Extrusion is a failure mode considered for untreated soil remaining between treated soil walls which is subjected to unbalanced active and passive earth pressure, as shown in Figure 5.10.

In the current design procedure for extrusion failure, soft soil between the long walls is assumed to move as a rigid body in the shape of a rectangular prism, where the width and length of the prism are taken as the width of the short wall and length of the long wall respectively (see Fig. 5.11). In the examination of extrusion failure, the minimum safety factor calculated by Eq (5.10) changing the height of the prism should be higher than the allowable value. Terashi et al. (1983) performed a series of centrifuge model tests of the failure pattern and proposed a simple calculation.

80 *The Deep Mixing Method – Principle, Design and Construction*

Figure 5.10. Deformation of clay ground between long walls in extrusion failure.

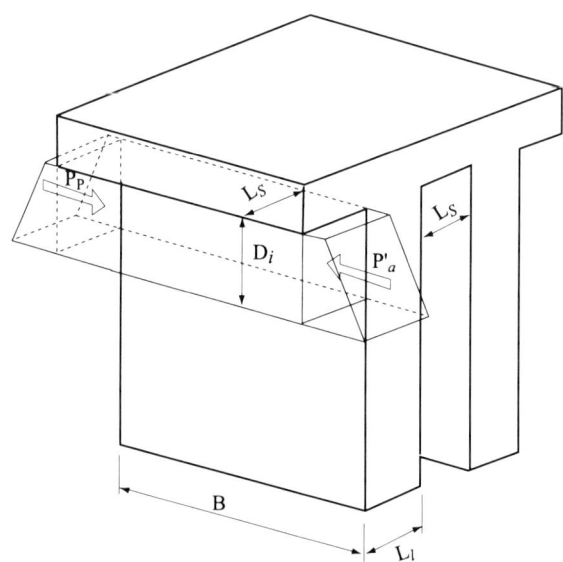

Figure 5.11. Conceptual diagram of the extrusion of unimproved soil.

$$Fs = \frac{2(Ls + Di) \cdot B \cdot cu + Pp'}{Pa' + kh \cdot \gamma \cdot B \cdot Di \cdot Ls + hw \cdot \gamma_w \cdot Di \cdot Ls} \quad \cdots\cdots \text{Eq. (5.10)}$$

where
- Di : height of prism (m)
- cu : mean undrained shear strength of the untreated soil

γ : unit weight of treated soil
kh : design seismic coefficient
hw : residual water level
γ_w : unit weight of water
Pa': total of active earth pressure
Pp': total of passive earth pressure

(e) *examination of settlement*. After the optimum cross section of the improved treated soil mass is determined by the above procedure, the immediate and long-term displacements of the improved ground should be examined. Usually, the deformation of the treated soil itself can be negligible because of its high rigidity and the large consolidation yield pressure of treated soil. Therefore, the displacement of the improved ground is calculated as the deformation of the soft layers surrounding or underlying the treated soil mass.

In the case of more complicated improvement pattern of wall and lattice (grid), untreated soils are left between the treated soil masses. It is desirable to apply sophisticated 2-D or 3-D elasto-plastic FEM analyses to examine stresses developed in the improved ground and displacement of the improved ground.

(3) *sample calculation*. An example of calculations for the most common DMM application is shown in Figure 5.12 (Terashi et al., 1985). In this example, the superstructure is a revetment composed of a gravel mound and a concrete caisson supporting the earth pressure induced by backfill. The superstructure is to be constructed on a soft clay layer underlain by a reliable bearing stratum of dense sand as shown in the upper left corner of the figure.

The initial approximation of the width of the improved ground in this trial calculation is shown by two vertical dotted lines at both ends of gravel mound in the figure. To obtain the necessary safety factor, the width B of the improved ground is increased by increasing distances la and/or lb. The three curves in the figure denote: the minimum extent of the treated soil mass that satisfies the requirement for the sliding failure of the improved ground (Curve I), the induced shear stress at the toe of the buried structure (II), and the shear stress in the vertical plane in front of the superstructure (III). The arrow added to each curve shows the direction toward a higher safety factor for each mode of failure. The hatched zone in the figure satisfies all the requirements and the dimension at point "A" is the optimum one. In this particular example, overturning, bearing capacity, and extrusion failures are not the governing factors. As is shown, usually a couple of failure modes become critical factors in determining the shape and extent of the improved ground.

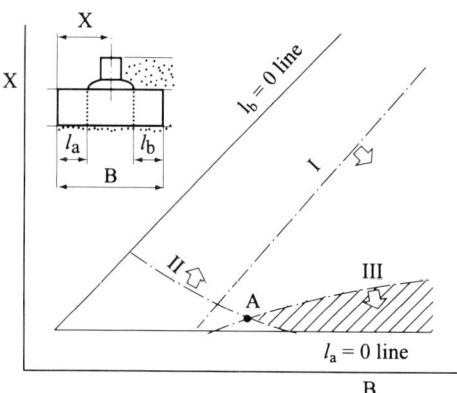

Figure 5.12. Determination of the optimum design (Terashi et. al., 1985).

5.3 DESIGN PROCEDURE FOR GROUP COLUMN-TYPE IMPROVEMENT

(1) *basic concept*. Group column-type DMM-improved ground has frequently been applied to embankments or retaining walls in order to prevent sliding failure, to reduce deformation and, to increase bearing capacity. In the current design in Japan, group column-type improved ground is considered to be a sort of composite ground with an average strength of DMM columns and surrounding untreated soil (Public Works Research Center, 1999).

Although the designed strength and improvement area ratio of columns are dependent upon the purpose of the improvement and the original ground condition, shear strength between 100 and 600 kN/ m^2 and an improvement area ratio between 30% and 50% have often been applied in the case history for the prevention of sliding failure and lateral deformation. For the improvement of retaining wall foundations and the lateral resistance of bridge abutment foundation piles, improvement area ratios of 60% or 78.5% have often been applied. Because the mechanical properties of a stabilized column are far from piles having uniform high strength, the ratio of the width of improved area and column length is usually set to larger than unity.

(2) *design procedure*. The design procedure for group column type is usually carried out following the design logic shown in Figure 5.13 (Public Works Research Center, 1999).

Design of Improved Ground by DMM 83

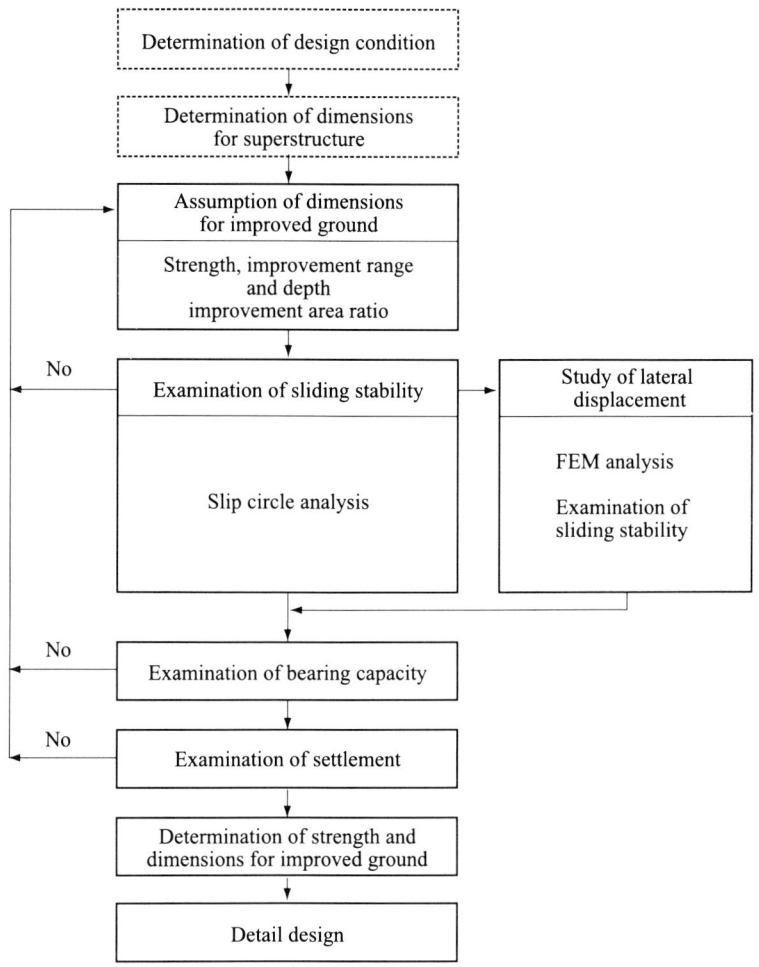

Figure 5.13. Flow chart of design for group column type improvement (Public Works Research Center, 1999).

The width and depth of the improved ground, the improvement area ratio and the strength of the column should be assumed at first by considering similar case histories. The improvement area ratio, *ap*, is, as shown in Figure 5.14, represented as the percentage of the sectional area of the column to the ground occupied by the column, and it is calculated by Equation (5.11). Generally, the DMM columns are arranged either in a rectangular or staggered pattern.

$$ap = \frac{Ap}{d_1 \times d_2} \qquad \text{Eq. (5.11)}$$

where

ap : improvement area ratio (%)
Ap : sectional area of DMM column
d_1 : intervals between columns
d_2 : intervals between columns

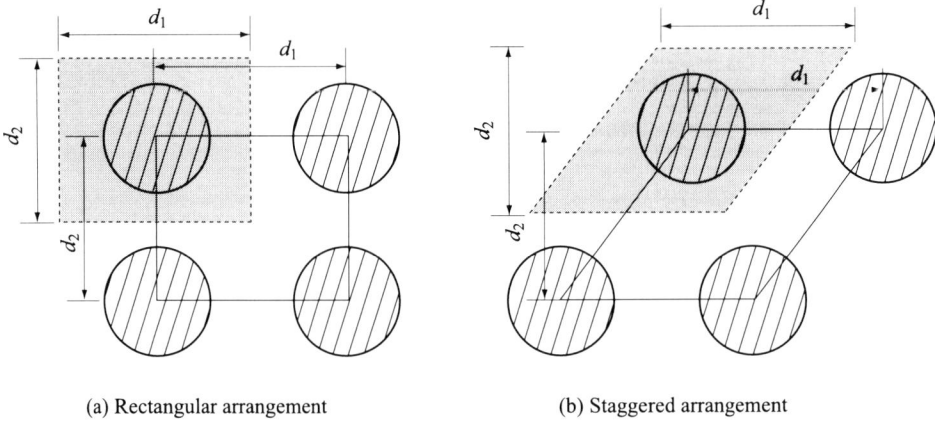

(a) Rectangular arrangement   (b) Staggered arrangement

Figure 5.14. Arrangement of stabilized columns.

The design unconfined compressive strength of a DMM column can be assumed at first by the following equation with the safety factor of 1.0 to 1.2.

$$qu_{ck} = Fs \frac{\gamma \times H}{ap} \quad \text{Eq. (5.12)}$$

where
    ap : improvement area ratio (%)
    qu_{ck} : sectional area of DMM column
    γ : density of embankment
    H : height of embankment
    Fs: safety factor

The strength of the improved body is from 10 to 100 times as high as the natural ground, and the compressive stress of the improved body is rarely a problem under the load of an embankment, but it is included in the basic study in cases where the improvement area ratio will be low or the embankment will be high. (See Fig. 5.15).

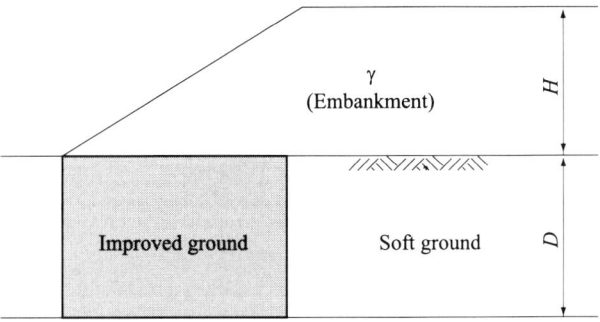

Figure 5.15. Improved strength and embankment height.

(a) *slip circle analysis*. The first stability step of the design procedure is slip circle analysis to determine the width of the improved ground. In the examination, the composite ground consisting of treated and untreated soils is assumed to have the average strength of the improved ground defined by Equation (5.13), as shown in Figure 5.14. The safety factor to slip circle failure is calculated by Equation (5.14) with Figure 5.16 (Public Works Research Center, 1999). It is easy to imagine from the equation that a small improved ground with very high strength is an alternative. Past experience, however, reveals that such an alternative is not suitable and the width of the improved ground should be larger than a half of the thickness of the soft ground in order to obtain stability in the ground.

$$\left. \begin{array}{l} \tau = ap \cdot Cp + (1-ap) \cdot Coo \\ Coo = k \cdot C \end{array} \right\} \quad \text{Eq. (5.13)}$$

where
- ap : improvement area ratio
- Cp : shear strength of the stabilized column
- Co : shear strength of the soft soil
- Coo: shear strength of soft soil mobilized at the peak of the shear strength of treated soil
- τ : average shear strength of improved ground
- k: coefficient factor for soft soil strength

86 *The Deep Mixing Method – Principle, Design and Construction*

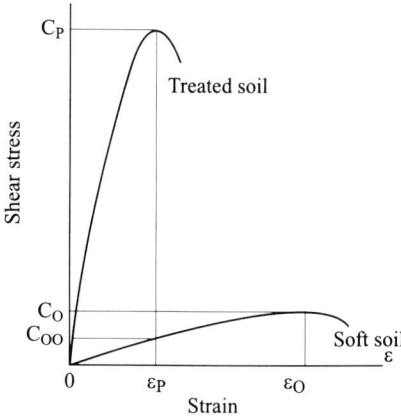

Figure 5.16. Determination of the average strength (Kitazume et al., 1996).

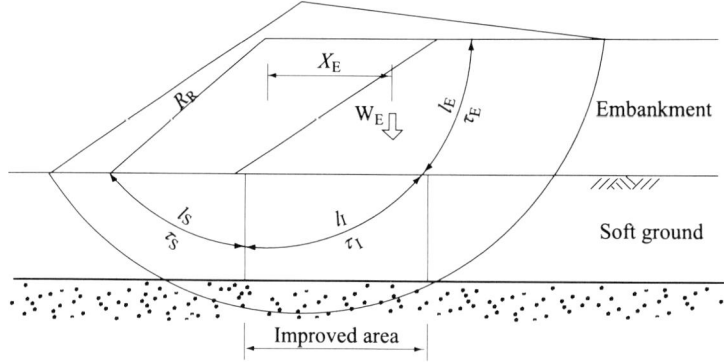

Figure 5.17. Slip circle analysis (Public Works Research Center, 1999).

$$Fs = R_R \cdot (l_E \cdot \tau_E + l_I \cdot \tau_I + l_S \cdot \tau_S) / X_E \cdot W_E \quad \cdots\cdots\cdots\cdots \text{Eq. (5.14)}$$

where
- l_E : length of circular arc in embankment
- l_I : length of circular arc in improved ground
- l_S : length of circular arc in soft ground
- R_F : radius of slip circle
- W_E : weight of embankment
- X_E : horizontal distance of embankment from center of slip circle
- τ_I: average shear strength of improved ground
- τ_E : shear strength of embankment
- τ_S : shear strength of soft ground

(b) *sliding failure analysis*. In the next step, the sliding failure of the improved ground is examined, in which the improved ground is assumed to behave as a whole. In the design, the stability is calculated based on the unbalanced earth pressure acting on both sides of the improved ground (Fig. 5.18). The safety factor against sliding failure is calculated by an equation similar to Equation (5.2).

$$Fss = \frac{Pp_S + Fr}{Pa_E + Pa_S}$$
$$= \frac{Pp_S + (W_E + W_I) \cdot \tan\phi'}{Pa_E + Pa_S} \quad \text{Eq. (5.15)}$$

where
- Fr : shear strength acting on the bottom of the improved ground
- Fss: safety factor against sliding failure
- Pa_E: active earth pressure of embankment
- Pa_S: active earth pressure of soft ground
- Pp_S: passive earth pressure of soft ground
- W_E: weight of embankment
- W_I: weight of improved ground
- ϕ': internal friction angle of stiff ground underlying soft ground

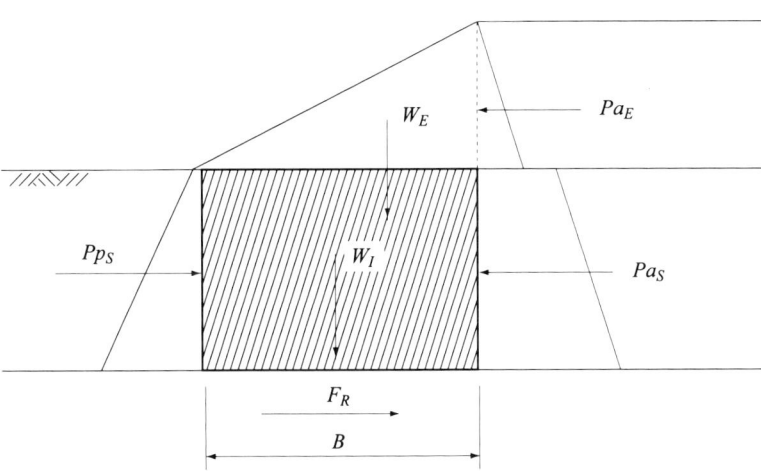

Figure 5.18. Sliding failure analysis.

(c) *bearing capacity analysis*. The bearing capacity of the improved ground is then examined. It is assumed in the calculation that the overburden pressure such as weight of embankment is concentrated on the DMM columns.

(d) *consolidation settlement*. The fourth step of the design procedure is the evaluation of the consolidation settlement. The amount of settlement of improved ground overlying a stiff layer can be obtained by the following equation, in which uniform settlement is assumed for the stabilized column and untreated soil (see Fig. 5.19 and Equation (5.16)).

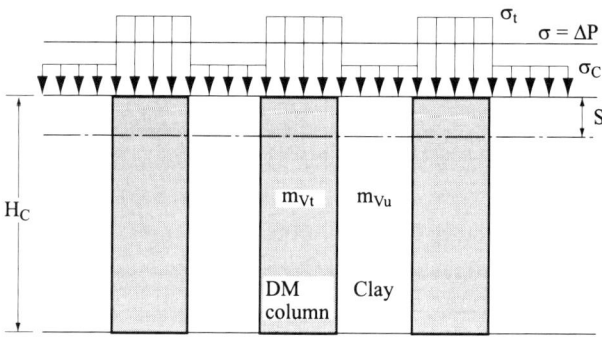

Figure 5.19. Calculation of consolidation settlement.

$$S = \beta \cdot S_0$$
$$\beta = \sigma_c / \sigma = \frac{1}{((n-1) \cdot ap + 1)}$$
$$S_0 = mvu \cdot Hc + \Delta P$$

Eq. (5.16)

where
- ap: improvement area ratio
- S: consolidation settlement of improved ground
- S_o: consolidation settlement of soft soil ground without improvement
- β: coefficient factor for stress reduction
- σ: increment of vertical stress
- σ_t: vertical stress acting on treated soil columns
- σ_c: vertical stress acting on soft soil between treated soil columns
- mvt: coefficient of volume compressibility of treated soil
- mvu: coefficient of volume compressibility of untreated soil
- n: stress concentration ratio ($= \sigma_c / \sigma_t = mvu / mvt$)
- Hc: thickness of the improved layer
- ΔP: increment of overburden pressure

The stress concentration ratio n, is usually set from 10 to 20 in practical design. In the case where a compressible layer is overlaid by the improved ground, settlement of this layer should be estimated adequately.

When the improvement area ratio of the improved ground, *ap* becomes about 78%, the improved ground is regarded as a columns in contact-type improvement in which the treated soil columns are in contact with each other. The columns in contact-type improvement has been employed to form the foundations of retaining walls to increase horizontal resistance. In the stability analysis, lateral displacement is studied based on the slip surface method and a safety factor larger than 1.5 is usually adopted to ensure relatively small displacement (Kitazume et al., 1991).

(4) *precautions in design*. In the slip circle analysis in the design procedure as explained in (2) design procedure, only shear failure is assumed as a failure mode of a DM column. Recent research efforts, however, reveal that the stabilized columns fail by several failure modes such as shear failure and bending failure (Kitazume et al., 2000). It is also evident that the improved ground also fails by several failure patterns including rupture breaking failure and collapse (see Appendix C). Appropriate design should therefore be adopted by considering the strength and settlement properties of improved ground, and accumulated case histories. If the improvement area ratio, *ap*, becomes less than 0.3 and/or the width of the improved ground becomes small, another failure pattern assumed in the current design might occur such as extrusion failure of untreated soil or collapse failure of the column (Figs 5.20 - 5.22).

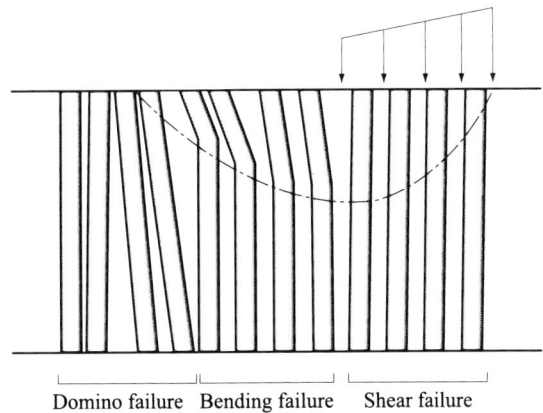

Figure 5.20. Forms of failure in DMM-improved ground.

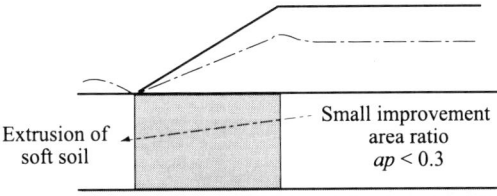

Figure 5.21. Failure patterns in cases of a small improvement area ratio.

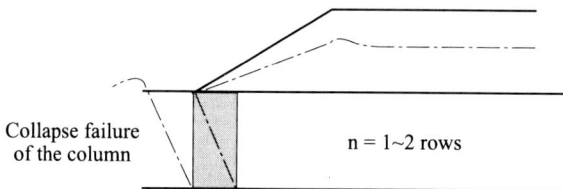

Figure 5.22. Failure pattern in case of small improvement width.

References

Cement Deep Mixing Method Association. 1993. *Cement Deep Mixing Method Manual*: 192 (in Japanese).

Karastanev, D., M. Kitazume, S. Miyajima, & T. Ikeda. 1997. Bearing capacity of shallow foundations on column-type DMM-improved ground. *Proc. of 14th International Conference on Soil Mechanics and Foundation Engineering*, 13: 1621 – 1624.

Kawasaki, T., A. Niina, S. Saitoh, Y. Suzuki & Y. Honjyo. 1981. Deep Mixing Method using cement hardening agent. *Proc. of the 10th International Conference on Soil Mechanics and Foundation Engineering*, 3: 721-724.

Kitazume M. 1994. Model and Analytical Studies on Stability of Improved Ground by Deep Mixing Method. *Technical Note of the Port and Harbour Research Institute*, 774: 73 (in Japanese).

Kitazume, M., T. Nakamura & M. Terashi. 1991. Reliability of clay ground improvement by group column-type DMM with high replacement. *Report of the Port and Harbour Research Institute*, 30(2): 305 – 326 (in Japanese).

Kitazume M., K. Omine, M. Miyake & H. Fujisawa. 1996. Japanese Geotechnical Society Technical Committee Report – Japanese design procedures and recent DMM activities, Grouting and Deep Mixing. *Proc. of the Second International Conference on Ground Improvement Geosystems*, 2: 925 – 930. Balkema.

Kitazume, M., K. Okano & S. Miyajima. 2000. Centrifuge model tests on failure envelope of column-type DMM-improved ground. *Soils and Foundations*, 40(4): 43-55.

Ministry of Transport. 1999. Design codes for port and harbour facilities (in Japanese).

Nakamura, R. 1997. Deep Mixing Method using cement slurry as hardening agent. *Umetate To Shunzetsu*, 78: 51 (in Japanese).

Noto, S., N. Kuchida & M. Terashi. 1983. Actual practice and problems on the deep mixing method. *Proc. of the Journal of Japanese Society of Soil Mechanics and Foundation Engineering, Tsuchi To Kiso*, 31(7): 73-80 (in Japanese).

Public Works Research Center. 1999. *Deep mixing method design and execution manual for land works*: 92-99 (in Japanese).

Sugiyama, K. et al. 1980. Soil Improvement Method of Marine Soft Soil by Cement Stabilizer. *Doboku Sekou*, 21(5): 65-74 (in Japanese).

Terashi M., & M. Kitazume. 1987. Interference effect on bearing capacity of foundations on sand. *Report of the Port and Harbour Research Institute*, 26(2): 413 – 436 (in Japanese).

Terashi M., M. Kitazume & T. Nakamura. 1989. External forces acting on a stiff soil mass improved by DMM. *Report of the Port and Harbour Research Institute*, 27(2): 147-184 (in Japanese).

Terashi M., M. Kitazume & M. Yajima. 1985. Interaction of soil and buried rigid structures. *Proc. of the 11th International Conference on Soil Mechanics and Foundation Engineering*, 13: 1757 – 1760.

Terashi, M., H. Tanaka & M. Kitazume. 1983. Extrusion failure of ground improved by the deep mixing method. *Proc. of the 7th Asian Regional Conference on Soil Mechanics and Foundation Engineering*, 1: 313-318.

Terashi, M., H. Tanaka & T. Okumura. 1979. Engineering properties of lime treated marine soils and the Deep Mixing Method. *Proc. of the 6th Asian Regional Conference on Soil Mechanics and Foundation Engineering*: 191-194.

Terashi, M., H. Tanaka, T. Mitsumoto, Y. Niidome & S. Honma. 1980. Fundamental properties of lime and cement treated soils (2nd Report). *Report of the Port and Harbour Research Institute*, 19(1): 33-62 (in Japanese).

Terashi, M. & H. Tanaka. 1981. Ground Improved by the Deep Mixing Method. *Proc. of the 10th International Conference on Soil Mechanics and Foundation Engineering*, 3: 777-780.

Terashi M., H. Tanaka & M. Kitazume. 1983. Extrusion failure of ground improved by the deep mixing method. *Proc. of the 7th Asian Regional Conference on Soil Mechanics and Foundation Engineering*, 1: 313 – 318.

CHAPTER 6

Construction Procedures and Control

6.1 CLASSIFICATION OF TECHNIQUES

The techniques for the Deep Mixing (DM) method can be divided into two groups: mechanical mixing and the high-pressure injection. The various methods in these groups are classified in Figure 6.1, in which the most common methods used in Japan are included. In the mechanical mixing technique, a stabilizing agent is fed to soft ground and forcibly mixed with the soft soil by mixing blades. The stabilizing agent is used either with a slurry form or as a dry form. The Cement Deep Mixing (CDM) method, the most common slurry type technique, has frequently been applied for both marine and land construction (Cement Deep Mixing Method Association, 1993). The Dry Jet Mixing (DJM) method is the most common powdery type technique and is usually applied for land construction (Dry Jet Mixing Association, 1993). In the high-pressure injection technique, on the other hand, the original ground is softened with high-pressure jet of water and/or air, while at the same time a stabilizing agent is introduced and mixed with the soil. Recently, a new method has been developed which exploits the features of both the basic techniques, combining mechanical mixing and high-pressure injection (Endo, 1995).

In this chapter, execution technique and execution control for the mechanical mixing technique is described.

6.2 MARINE WORKS

(1) *DM machine*. In practice, a special barge equipped with a DM machine is used to improve soft soil in situ (Fig. 6.2). The DM machine must have the capability of uniformly supplying the soft soil with a stabilizing agent and of achieving a sufficient mixedness of stabilizing agent and soft soil.

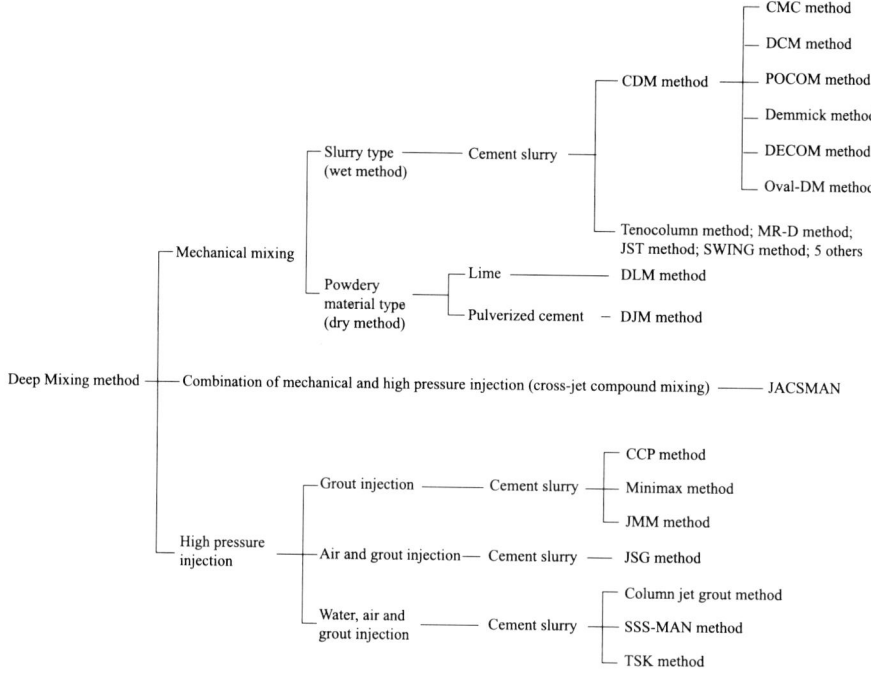

Figure 6.1. Classification of Deep Mixing Methods.

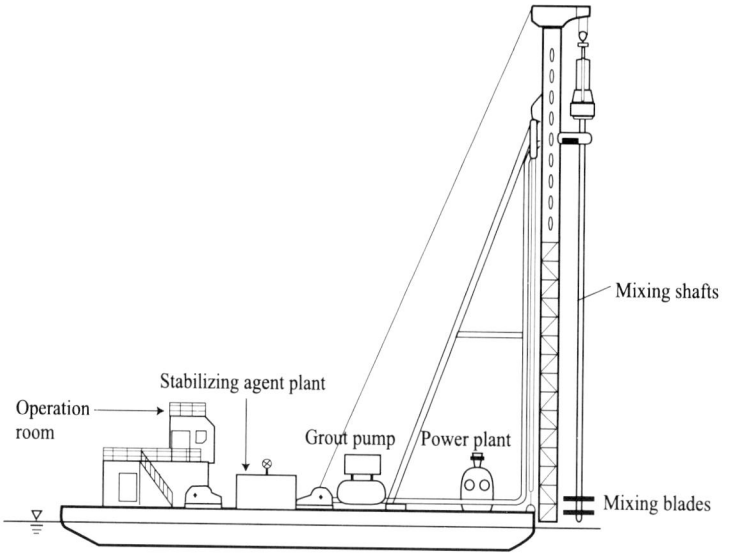

Figure 6.2. DM barge for marine work.

DM machines for marine work usually have more than two mixing shafts. Shown in Figure 6.3 is an example of the bottom end of a DM machine that has eight mixing shafts. The DM machines currently available in Japan are capable of constructing large columns whose cross-sectional area ranges from 1.5 m^2 to 9.5 m^2 and maximum depth of stabilizing reaches up to 70 m from the water surface (Fig. 6.4). A DM machine can penetrate local stiff layers to reach the desired depth. A machine with a relatively large capacity can penetrate a layer whose SPT N-value and thickness are 8 and 4 m for clayey soil, and 15 and 4 m for sandy soil, respectively.

Figure 6.3. Mixing blades of DM machine for marine work.

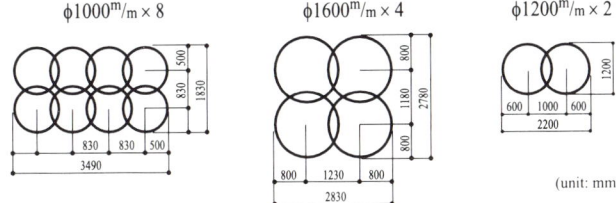

Figure 6.4. Typical column arrangements for marine work.

(2) *construction procedure*. A typical construction procedure for marine work is shown in Figure 6.5.

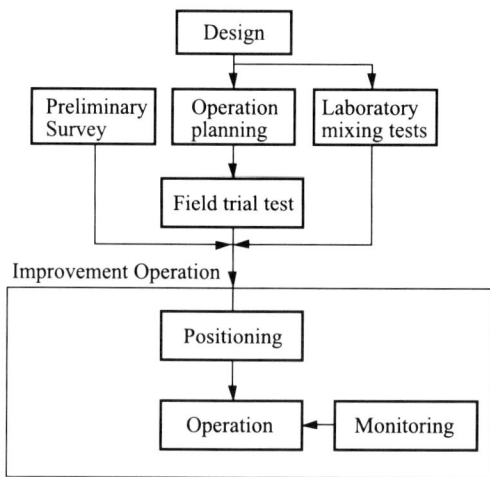

Figure 6.5. Typical DM method construction procedure for marine work.

a) *preliminary survey*. Before actually executing ground improvement, execution circumstances should be checked to ensure sufficient quality of the improved ground, smooth operation and avoidance of environmental impact. The execution schedule can be disturbed by severe weather and wave conditions. Thus weather and wave conditions should be examined in advance when making the execution schedule; wave height, wind direction, wind velocity and tides should be carefully surveyed. According to past records, marine work is difficult to conduct in conditions where the maximum wind velocity exceeds 10 m/second, the maximum significant wave height exceeds 50 cm, or the minimum visibility is less than 1,000 m. Environmental impacts such as water contamination, noise, vibration etc. which can occur during the execution should obviously be kept to a minimum.

In the DM method, the mixing blades at the bottom end of the mixing shaft rotate during penetration into soft ground. Any obstacles on or below the seabed in the construction area can thereby delay the operation schedule, or cause damage to the blades. Before operations, the seabed should be surveyed carefully and any obstacles should be removed. This process is particularly important with regard to blind shells that can cause human damage. This soil survey can usually be carried out by means of magnetic prospecting probe.

b) *field trial test*. It is advisable to conduct the field trial test in advance at a ground in or adjacent to the construction site, in order to ensure the smooth execution at a specific site. In the test, all the monitoring equipment, such as amount of stabilizing agent, rotation speed of mixing blade and penetration and withdrawal speeds of shaft are calibrated. In addition, the field trial test can be useful to confirm whether the design mixing condition will be adequate when the strength of the treated soil is measured after the trial test.

c) *improvement operation (positioning/ processing)*. The positioning methods of the DM machine for marine work have four alternatives; collimation of two transit apparatuses, collimation using a transit and an optical range finder, an automatic positioning system with three optical finders, and a positioning system with GPS.

A column of treated soil is usually manufactured by the procedure shown in Figure 6.6. During the penetration of the mixing shaft to the desired depth of improvement, the mixing blades at the bottom end of the mixing shafts cut and disturb the soft soil to reduce the strength of the original soil. Then the machine can sink into the soft ground by its own weight. In the withdrawal stage of the machine, the vertical speed of the machine is kept constant. At the same time, the stabilizing agent in a slurry form is injected into the soft soil at a constant flow rate. The mixing blades rotate in the horizontal plane and mix the soft soil and the stabilizing agent. Thus a column of soil-stabilizing agent mixture having a cross section as shown in Figure 6.4 is manufactured in situ. There are some variations of machine, in which the stabilizing agent is injected during the penetration stage and mixing with the original ground is done both in the penetration and withdrawal stages.

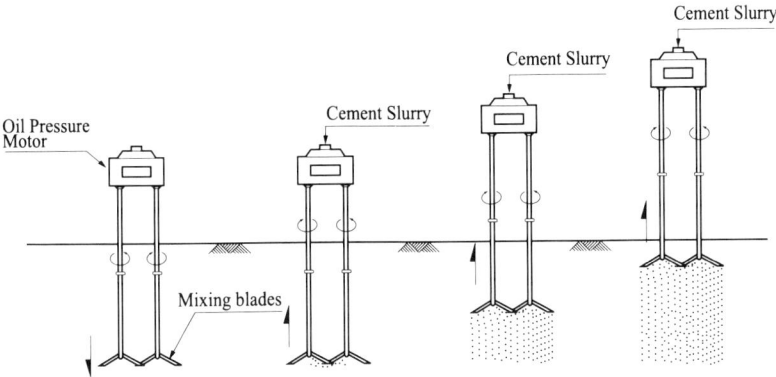

Figure 6.6. Procedure of manufacturing a treated soil column in-situ.

An air entraining (AE) agent or water reducing agent is often used together with the stabilizing agent to improve the fluidity of the slurry and/or to control the rate of pozzolanic reaction between the stabilizing agent and soil. While manufacturing the stabilized columns in situ, the amount of stabilizing agent, rotation speed of mixing blades, and penetration and withdrawal speeds of the shaft are continuously monitored. These data are displayed in the operation room for the operator usually at intervals of 1 m shaft movement to control the shaft movement speed and assure the designed amount of stabilizing agent injected. The degree of mixing effectiveness mostly depends on the rotation speed of the mixing blade and penetration and withdrawal speed of the shaft. In Japan, an index named "blade rotation number", T has been introduced to evaluate the mixing degree. This number means the total number of mixing blade passes during 1 m of shaft movement and is defined by the following equation for the possible penetration injection method. There are, of course, various combinations of rotation speed and shaft speed possible to obtain the prescribed "blade rotation number". The following combination however, is usually adopted in the field execution based on many research works and field experiments in Japan. Some of them are briefly described in Appendix B. Typical penetration and withdrawal speeds of the machine are about 0.3 m to 1.0 m/min, and the rotation speed of the blade is about 20 to 100 rpm.

$$T = \sum M \times \{(Nd/Vd) + (Nu/Vu)\} \qquad \text{Eq. (6.1)}$$

where
 T : blade rotation number (n/m)
 $\sum M$: total number of mixing blades
 Nd : rotation speed of the blades during penetration (rpm)
 Vd : mixing blade penetration velocity (m/min)
 Nu : rotational speed of the blades during withdrawal (rpm)
 Vu : mixing blade withdrawal velocity (m/min)

There are two basic execution procedures depending on the injection sequence of stabilizing agent (Fig. 6.7): (a) injecting the stabilizing agent during penetration of the mixing machine and (b) injecting the stabilizing agent during withdrawal of the mixing machine. Each injection sequence has its respective advantages and disadvantages. The penetration injection method is beneficial for the homogeneity of the column strength in which the original soil is twice subjected to mixing with a stabilizing agent. However, it is possible to deadlock or cause serious damage to the machine if any trouble occurs with the mixing machine during penetration. The withdrawal injection method has the opposite benefits and disadvantages to the penetration injection method. In Japan, the penetration injection method is frequently applied for the DJM method in land construction. The withdrawal injection method is frequently applied to the CDM method, in which sufficient mixture is achieved during the withdrawal stage to assure the homogeneity of the stabilizing column. The location of the injection outlet is slightly different for each injection method to assure greater homogeneity of the column. For the penetration injection method, the injection outlet is placed close to the bottom end of the mixing shaft, but it is above the mixing blades for the withdrawal injection.

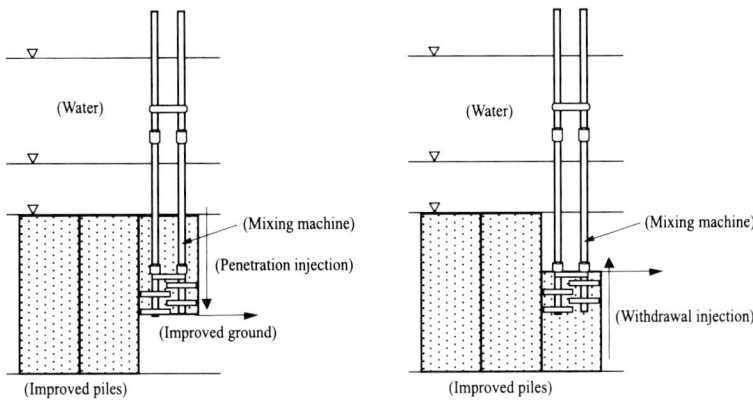

(a) penetration injection method   (b) withdrawal injection method

Figure 6.7. Basic execution procedure.

To assure the external stability of improved ground, the stabilized columns should reach the stiff layer and have sufficient contact in the case of the fixed- type improvement. It is common that the depth of the stiff layer has local undulation and is different from the design value determined by the soil survey. Therefore it is essential in the actual execution to confirm the depth of the stiff layer for every column and to adjust the column length based on at the depth of the stiff layer. In the execution, the rapid change in penetration velocity of the shaft, required torque and rotation speed of the mixing blade are key indicators to determine whether the machine has reached the stiff layer. After reaching the stiff layer, the mixing blades should go up and down several times together with injection to assure the sufficient contact of the column and the base layer.

d) *operation monitoring (quality control/quantity control)*. To assure the quality and dimension of the stabilized column, it is essential to keep the designed condition in the actual execution by monitoring the admixture condition, quantity each material, rotation speed of mixing blade, speed of shaft movement, etc. These monitoring data can be fed back to the operator for precise construction of the column. As shown in Figure 6.8, the operation monitoring used in the DM method covers quality and quantity control monitoring. The gauges and meters normally used are marked ① to ⑧ in Figure 6.8. The cement slurry flow gauges, the vertical speed gauge and depth gauge are absolutely essential.

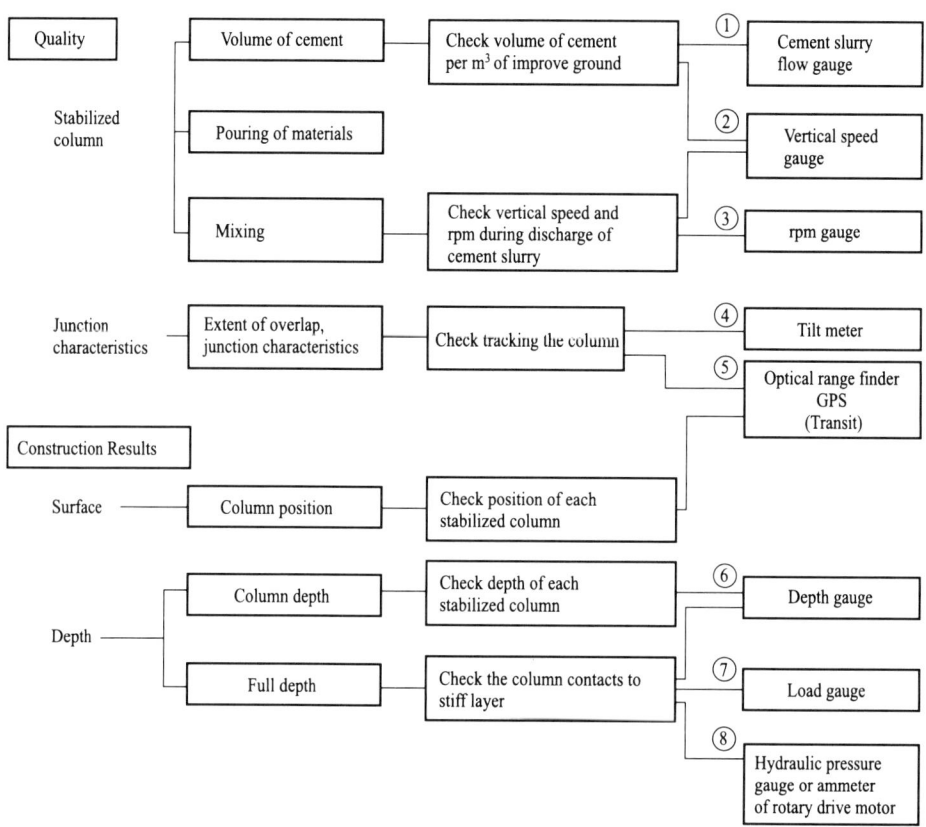

Figure 6.8. Monitoring chart for marine works.

e) *heaving of sea bottom*. As a result of injecting stabilizing agent into a soft ground, the ground surface can heave to some extent. Figure 6.9 shows a typical case record on upheaved ground at Yokohama Port in which 160 kg/m^3 of cement slurry (w/c of 0.6) was injected into the ground. The extent of the upheaved ground is not uniform and will depend on many factors such as the soil profile, the thickness of the improved layer, the improvement area ratio, and the workflow sequence. According to accumulated field experience, the total volume of upheaved soil is almost equivalent to that of the cement slurry injected, and the upheaved volume within the improved ground area is approximately 70% of the volume of the cement slurry injected.

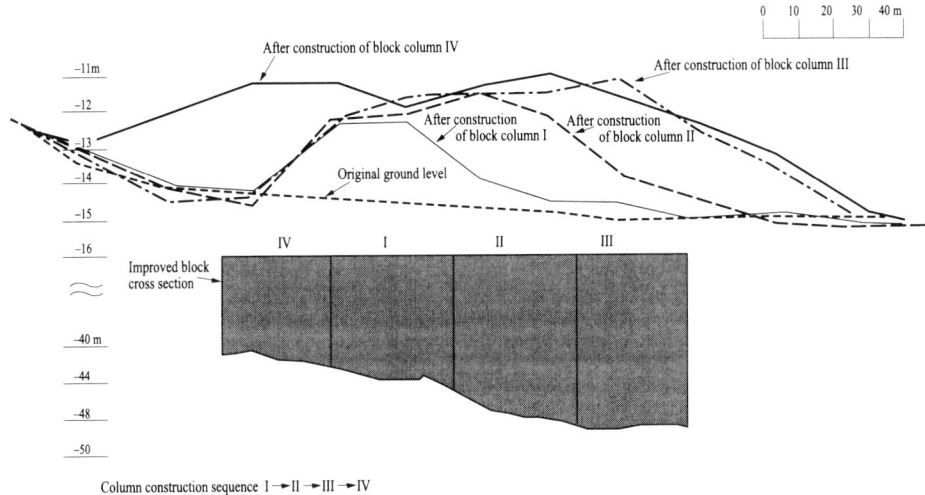

Figure 6.9. A typical case record on upheaved ground at Yokohama Port.

Since the upheaved soil is usually disturbed and softened, it is usually handled by one of the following procedures: i) dredge and dispose of the soil to the determined depth; ii) improve to a level close to the surface, then dredge and dispose of the upper surface layer; iii) improve the soil to the surface. The first procedure has been used in most cases in Japan to obtain the required water depth for quay structures.

6.3 ON LAND WORKS

(1) *DM machine.* DM machines (Fig. 6.10) for on land work usually have one or two mixing shafts (Fig. 6.11). The mixing blade has a diameter of about 1 m, which can make a column with a cross-sectional area ranging from 0.8 m^2 to 1.5 m^2. The maximum depth of treatment reaches down to 40 m. The two shaft-type mixing machine usually has a bracing plate to keep the distance of the two mixing shafts (see Fig. 6.11). The plate is also expected to function to increase the mixing degree by preventing the "entrained rotation phenomenon", a condition in which disturbed soft soil adheres to and rotates with the mixing blade without efficient mixing of the cement slurry and soil. For the single blade type mixing head, a "free blade", an extra blade about 10 cm longer than the mixing blades (see Fig. 6.11), is usually installed close to the mixing blades to prevent the "entrained rotation phenomenon".

102 *The Deep Mixing Method – Principle, Design and Construction*

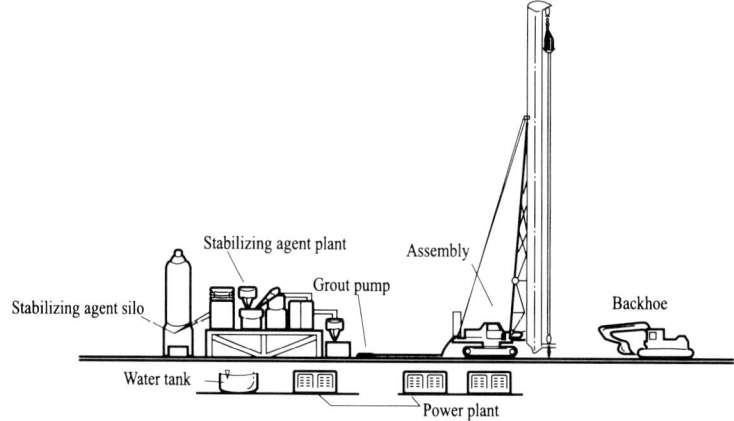

Figure 6.10. Typical DM machine for on land work.

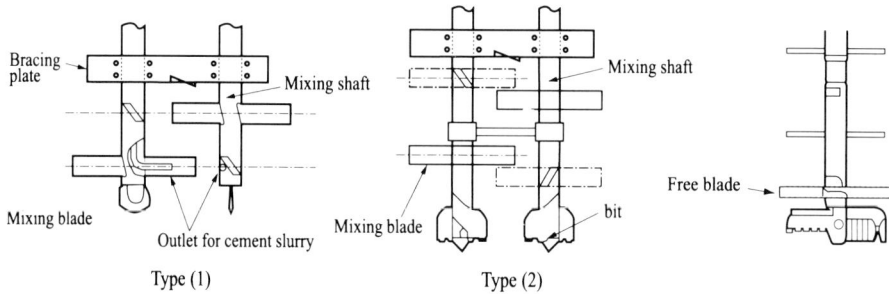

Figure 6.11. Mixing blades of DM machine for on land work.

(2) *construction procedure*

a) *preparatory survey*. A typical construction procedure for on land work is basically the same as that for marine work as already shown in Figure 6.5. Before actual operation, execution circumstances should be checked to assure sufficient quality of the improved ground, smooth operation and avoidance of environmental impact.

b) *field trial test*. It is advisable to conduct the field trial test in advance at a ground in, or adjacent to, the construction site, in order to confirm the smooth execution at a specific site. In the test, all the monitoring equipment such as amount of stabilizing agent, rotation speed of mixing blade and penetration and withdrawal speeds of shaft are calibrated. And the field trial test can be useful to confirm that the designed mixing condition is adequate if the strength of the treated soil is measured after the trial test.

c) *improvement operation*. A column of treated soil is usually manufactured by the procedure as shown in Figure 6.12. During the penetration of the mixing shaft to the desired depth of improvement, mixing blades at the bottom end of the mixing shafts cut and disturb the soft soil to reduce the strength of the original soil. At the same time, the stabilizing agent is forced into the soft soil at a constant flow rate. As described in Section 6.1, the stabilizing agent in a slurry form is used in CDM and in a dry form in DJM. In the withdrawal stage of the machine, the mixing blades rotating reversibly in the horizontal plane mix the soft soil and the stabilizer again. Typical penetration and withdrawal speed of the machine is about 1.0 m/min, and the rotation speed of the mixing blade is about 20 rpm and 40 rpm during the penetration and withdrawal stage respectively. This corresponds to a "blade rotation number" of about 360. These numbers are provided to assure sufficient homogeneity of the stabilized column according to many research efforts. Some of the research results on the effect of mixing conditions on the column strength are briefly described in Appendix B.

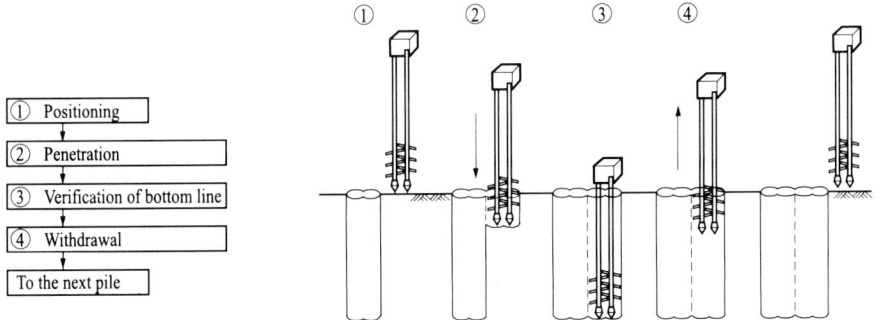

Figure 6.12. Procedure of the DM method for on land work.

Similarly to the marine construction, the stabilized columns should reach the stiff layer sufficiently in the case of a fixed-type improvement. In practical execution, the rapid change in penetration speed of the shaft, required torque and rotation speed of mixing blade are useful to determine whether the machine has reached the stiff layer. After doing so, the machine stays there for several minutes with continuing injection to assure sufficient contact with the stiff layer.

d) *operation monitoring (quality control/quantity control)*. To assure the quality and dimension of stabilized column, it is essential to keep the designed condition by monitoring the admixture condition, quantity of each material, rotation speed of mixing blade, shaft speed, etc. These monitoring data can be fed back to the operator for precise construction of the column. In practice, penetration and withdrawal speeds are commonly adjusted to feed the designed amount of stabilizing agent into the ground.

e) *additional considerations in execution*

- *Soil improvement method for locally hard ground*. Since the layer system in on land is usually complicated by the past geological history, it is not unusual to encounter a local stiff layer before reaching the desired depth. The DM machine for on land construction is usually smaller in size than that for marine construction. In some cases, the mixing blades and shaft of the DM machine may be damaged and/or stuck in the ground. When penetrating a hard layer, therefore, it is necessary to carry out pre-boring with an auger machine, or use a large powerful machine.

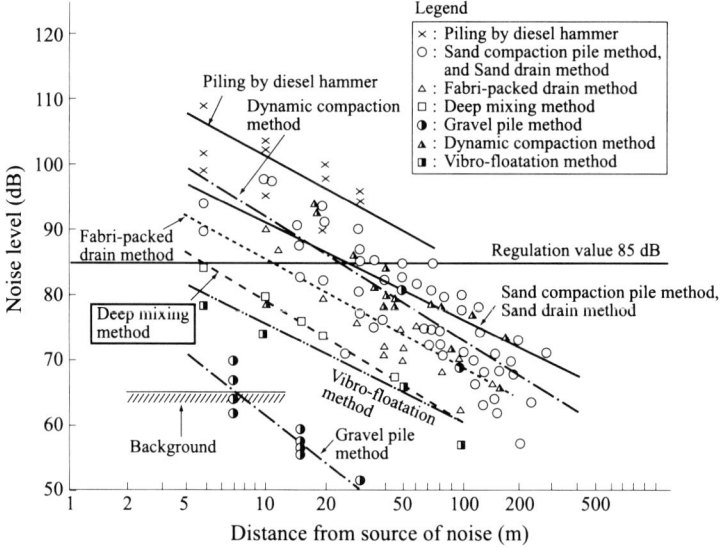

(a) Noise level and distance from the source

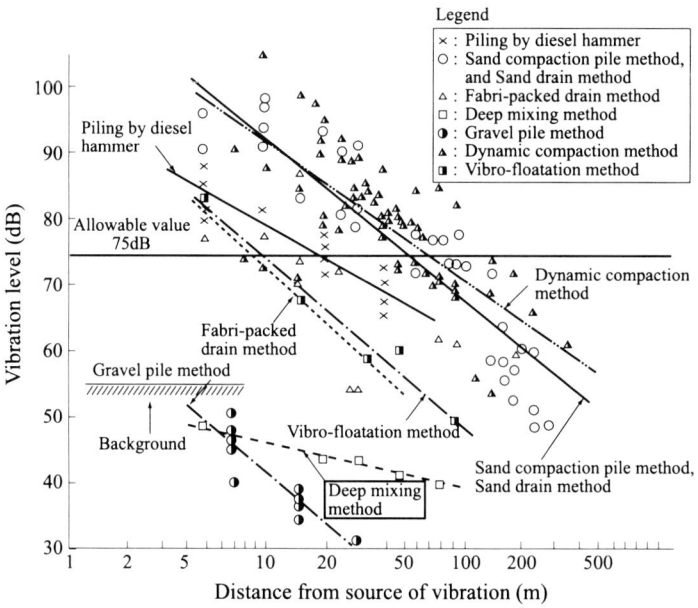

(b) Vibration level and distance from the source

Figure 6.13. Noise and vibration during the operation (Japanese Society of Soil Mechanics and Foundation Engineering, 1985).

- *Noise and vibration during the operation*. Figure 6.13 shows the relationship between noise and vibration levels and the distance from the source, in which the field values caused by various ground improvement techniques are also presented for comparison (Japanese Society of Soil Mechanics and Foundation Engineering, 1985). The figure indicates that the noise and vibration levels caused by the Deep Mixing Method are relatively small among the different soil improvement techniques and they are within the Japanese regulation values except at a point very close to the source.

- *Influence on surrounding ground*. As a result of injecting stabilizing agent into soft ground, the ground can heave to some extent and expand horizontally. Figure 6.14 shows examples of measured lateral displacement at the ground surface in which various topographic grounds and improved ground patterns are plotted. Although the amount of ground movement is relatively small compared with that in the marine construction (see Fig. 6.9) due to the small improvement area ratio, the ground moves horizontally 10 cm to 40 cm. It is important to estimate the amount of lateral displacement and its influence on the surrounding structure especially during construction in an urban area.

In recent years, a new type of DM method, the low displacement deep mixing method (LODIC method), has been developed for minimizing the lateral displacement during construction (Horikiri et al., 1996, Fig. 6.15). In this method an earth auger screw is installed on the upper part of the mixing shaft to remove the soil to the ground surface. If soil equivalent to the amount of cement slurry injected can be removed, it is basically possible to reduce the displacement of the surrounding ground or nearby structures substantially. Figure 6.16 shows the case record on the horizontal displacement during the improvement operation. In the figure, two case records by ordinary DM machine and the LODIC method are plotted. It is obvious that the LODIC method can reduce the horizontal movement considerably.

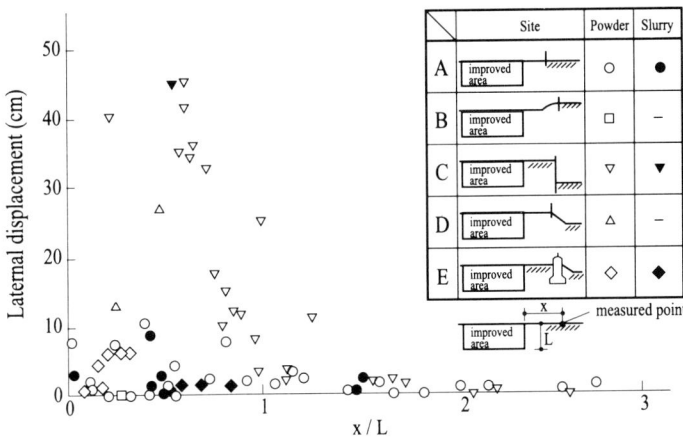

Figure 6.14. Lateral displacement of surrounding ground during improvement operation.

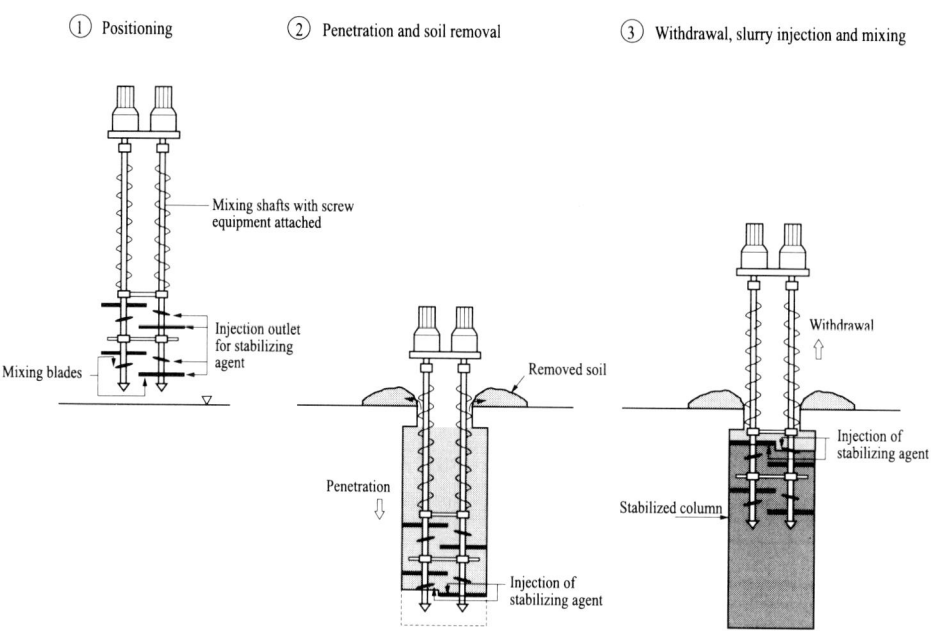

Figure 6.15. Soil improvement sequence in LODIC Method (Horikiri et al., 1996).

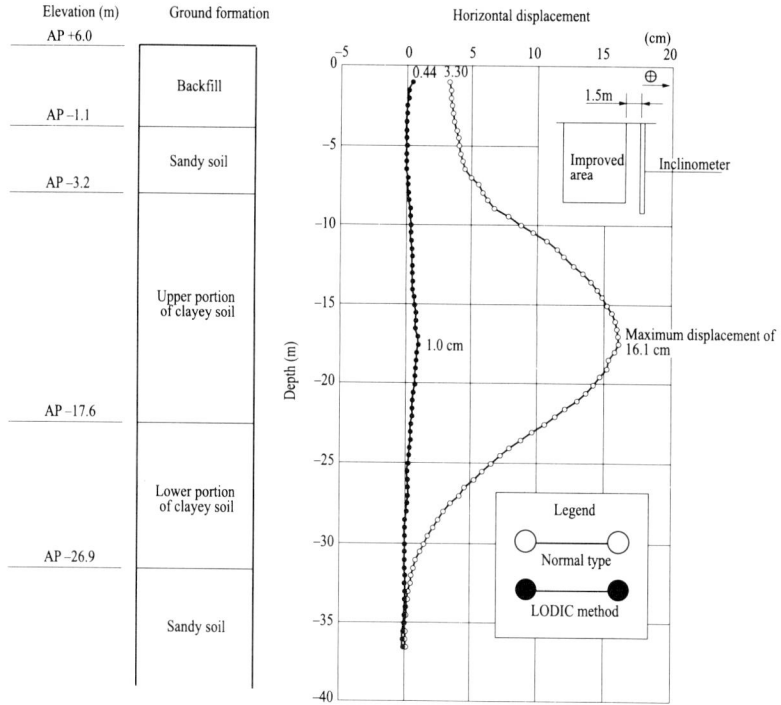

Figure 6.16. Measured horizontal displacement during improvement operation (Horikiri et al., 1996).

6.4 QUALITY CONTROL AND QUALITY ASSURANCE

(1) *viewpoint of quality control*. The design, construction and quality control procedure for the Deep Mixing method (DMM) is shown in Figure 6.17. To ensure the sufficient quality of the stabilized column, quality control and quality assurance is required before, during and after construction. For this purpose, quality control for the Deep Mixing Method mainly consists of i) laboratory mixing tests, ii) quality control during construction and iii) post construction quality verification through check boring and pile head inspection, as shown in Figure 6.17.

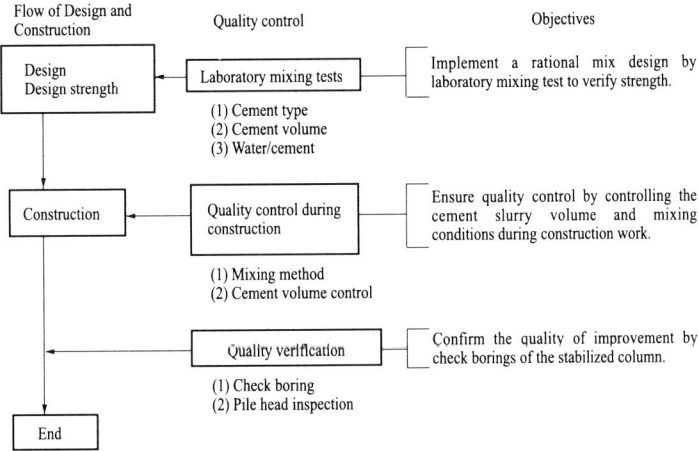

Figure 6.17. Flow chart for quality control and quality assurance.

As described in Chapter 2, the improvement effect achieved by the DMM is affected by many factors such as soil properties (natural water content, liquid limit, plastic limit, pH, organic matter content, mechanical components and clay minerals), type and quantity of stabilizing agent, degree of mixing, and curing conditions. The effects of these factors are quite complex, making it difficult to directly determine field strength by a laboratory mixing test.

The DM machines must be simple and tough enough to endure severe working conditions. Mixing time in practice must be as short as possible for economic reasons. These demands on the working machine unavoidably result in the reduction of mixedness in practical applications. Hence, in-situ mixing conditions and curing conditions are quite different from the standard laboratory testing, the strength of the in-situ stabilized column is usually different from that in the laboratory. And it is usual that the in-situ column has a relatively large scatter in strength even if the execution is done with a well established mixing machine and with the best care. Average compressive strength and the deviation in the laboratory specimen and in-situ column are schematically shown in Figure 6.18. It is usual that the in-situ column has smaller average strength and larger strength deviation than those of the laboratory specimen. In the figure, the in-situ designed strength, qu_{ck}, is derived from qu_f by incorporating the strength deviation. The target strength of the laboratory specimen should be determined by incorporating the strength difference and the strength deviation. When using statistical measures for quality control, the following relationship between in field strength and the design standard strength must be formulated if the field strength of the improved soil is assumed to have a normal curve.

$$\left. \begin{array}{l} qu_{ck} \leqq \overline{qu_f} - K \times \sigma \\ \overline{qu_f} = \lambda \cdot \overline{qu_l} \end{array} \right\} \quad \cdots\cdots\cdots\cdots\cdots\cdots \text{Eq. (6.2)}$$

where
$\quad qu_{ck}$: design standard strength (kN/m^2)
$\quad K$: coefficient
$\quad \sigma$: standard deviation of the field strength (kN/m^2)
$\quad \overline{qu_f}$: average unconfined compressive strength on in-situ stabilized column (kN/m^2)
$\quad \overline{qu_l}$: average unconfined compressive strength on laboratory treated soil (kN/m^2)

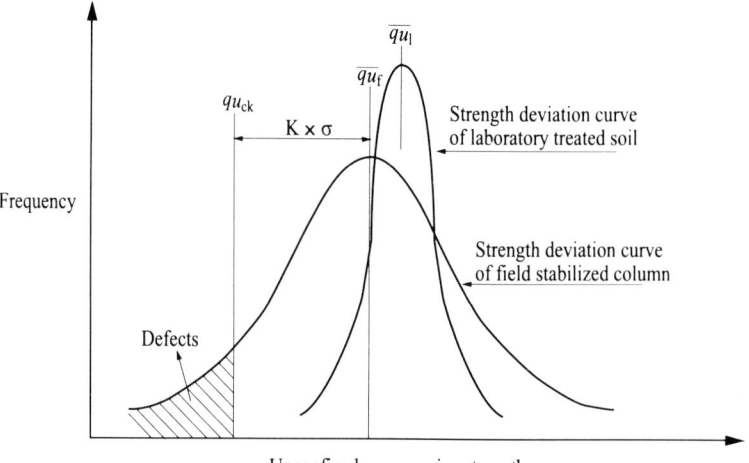

Figure 6.18. Determination of required laboratory strength.

(2) *laboratory mixing test.* Laboratory mixing tests should be conducted to determine a suitable type of stabilizing agent and a suitable quantity of the agent to ensure the design strength of the column in-situ. Portland cement or Portland blast furnace slag cement type B (including 30 to 60% slag) are usually used as a stabilizing agent in CDM and DJM. However, dozens of cement stabilizers are also available in the Japanese market for organic soils and extremely soft soil with

high water content (Japan Cement Association, 1994).

Laboratory strength is influenced by many factors, such as how to make the specimen, the curing condition, testing conditions. To avoid these factors, the Japanese Geotechnical Society standardized a laboratory mold test procedure in 1990, and made a minor revision in 2000 (Japanese Geotechnical Society, 2000). Almost all laboratory tests for practical and research purposes follow this standard in Japan, which helps us make direct comparisons of many factors. An outline of the standard is described in Appendix A.

(3) *quality control during construction*. During construction in-situ, amount of stabilizing agent, mixing rotation speeds of the mixing blade, penetration and withdrawal speeds of the shaft are usually monitored at the interval of 1 m shaft movement. These data are displayed in the operation room for the operator to adjust the execution procedure to ensure the quality. The degree of mixing effectiveness mostly depends on the rotation speed of the mixing blade and penetration and withdraw speeds of the shaft. "Blade rotation number" is introduced to evaluate the mixing degree.

(4) *post construction quality verification*. After the improvement operation, the quality of the in-situ stabilized columns should be verified in advance of the construction of the superstructure in order to confirm the design quality, such as strength, permeability or dimension. In Japan, unconfined compression tests on the stabilized column are most frequently conducted for quality verification; in which stabilized specimens are sampled from the construction site. The number of check borings is dependent upon the volume of the improved ground. In the case of on land construction, one check boring is generally conducted for every 3,000 m^3 of improved ground at 28 days of curing.

In the check boring on the stabilized column, DMM samples are taken throughout the depth in order to verify the continuity of the stabilized column. The quality of the sample relies on the quality of boring machine, sampling tool and the skill of a workman. Otherwise low quality samples with some cracks can be obtained. A Denison type sampler, double tube core sampler, or triple tube core sampler can be used for columns whose unconfined compressive strength ranges from 100 kN/m^2 to 6000 kN/m^2. It is advisable to use samplers of a relatively large diameter such as 86 mm in order to take high quality samples. An unconfined compression test is usually carried out on the samples, and the number of test depends upon the construction's condition and the soil properties. Recently, in-situ tests have also been applied for the quality verification together with the unconfined compression test; these include i) integrity test, ii) rotary sounding test and iii) vertical loading test.

References

Cement Deep Mixing Method Association. 1993. *Cement Deep Mixing Method (CDM) Design and Construction Manual* (in Japanese).

Dry Jet Mixing Association. 1993. *Dry Jet Mixing (DJM) Method Technical Manual* (in Japanese).

Endo, S. 1995. New large-diameter deep mixing method combining the advantages of mechanical mixing and jet stirring. *Proc. of the Journal of Japanese Society of Soil Mechanics and Foundation Engineering, Tsuchi to Kiso*, 448(5): 50 (in Japanese).

Horikiki, S., K. Kamimura & K. Kurinami. 1996. Low displacement deep mixing method (LODIC) and its application. *Kisokou* (24)7: 90-94 (in Japanese).

Japan Cement Association. 1994. *The manual of soil improvement using cement stabilizers* (2nd edition) (in Japanese).

Japanese Geotechnical Society. 2000. *Practice for making and curing stabilized soil specimens without compaction, JGS T 0821-2000*. Japanese Geotechnical Society (in Japanese).

Japanese Society of Soil Mechanics and Foundation Engineering. 1985. *Soil stabilization Techniques*: 389 (in Japanese).

APPENDIX A

Summary of the Practice for Making and Curing Stabilized Soil Specimens without Compaction (JGS 0821)

The method to prepare the stabilized soil specimen in a laboratory is standardized by the Japanese Geotechnical Society. This standard applies to a procedure of making and curing a cylindrical specimen of treated soil without compacting. An outline of the standard is described in the following paragraph.

For the laboratory mixing test, the original soil to be stabilized is sampled from the construction site. For ease of specimen preparation, the sampled soil is then sieved by a 9.5 mm sieve so that large-sized articles such as shells and/or plant roots can be removed. An electric mixer can be used to make mixture of the soil and stabilizing material (Fig. A.1). For the laboratory mixing test, either city (fresh) water or seawater may be used as mixing water, depending on the construction condition (on land work or marine work). In the mixing process, the original soil is mixed with the water first in the mixer to obtain the pre-determined water content. Then the stabilizing agent is added to the soil and mixed for about 10 minutes. The mixing time of 10 minutes is proposed based on the consideration of the well homogeneity and start time to harden. The soil mixture is put into the specimen mold with the diameter of 5 cm and the height of 10 cm. Its inside is lubricated with grease in advance. The soil mixture is divided into three layers to fill up the whole mold. After putting the mixture for the each layer, the mold is subjected to some vibration to remove air bubbles, which might be entrapped in the mixture. After covering the top surface of the specimen with a thin plastic sheet, the specimen is cured in the condition of a temperature of 20 ± 3 degrees Celsius and relative humidity of 95%. After some days' curing, the specimen can be removed from the mold and is cured under the same conditions again for the prescribed period.

112 *The Deep Mixing Method – Principle, Design and Construction*

Figure A.1. Electric mixer and mixing blade.

Reference

Japanese Geotechnical Society. 2000. *Practice for making and curing stabilized soil specimens without compaction, JGS T 0821-2000*. Japanese Geotechnical Society (in Japanese).

APPENDIX B

Influence of In-Situ Mixing Conditions on the Quality of Treated Soil

In Japan, there is a lot of research carried out to investigate several aspects in regard to operation techniques on the quality of treated soil. Some of the research results are described in the following paragraphs.

(1) *influence of number of mixing shafts*. The influence of number of mixing shafts on the strength of treated soil is shown in Figure B.1 (Nishibayashi et al., 1985). The figure shows a comparison of the results obtained with a single mixing shaft and a set of four mixing shafts. It is found the strength of the treated soil by a single mixing shaft is almost same order of those by four mixing shafts at 7 days, but it is much smaller at 28 days.

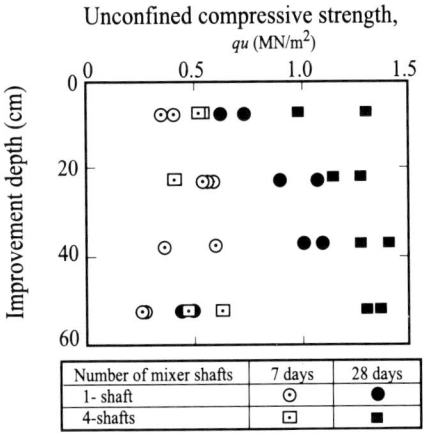

Figure B.1. Comparison of strength achieved using 1-shaft and 4-shafts model mixers (Nishibayashi et al., 1985).

(2) *influence of type and shape of mixing blade.* Figure B.2(a) shows the strength distribution with the depth, in which two types of mixing blade are used; open-type blade and horizontal-type blades (Abe et al., 1997) . Although there is a large scatter in the strength, the in-situ strength obtained by the open-type mixing blade is larger than those by the normal horizontal-type blade. This difference is dominant in the Humic soil layer.

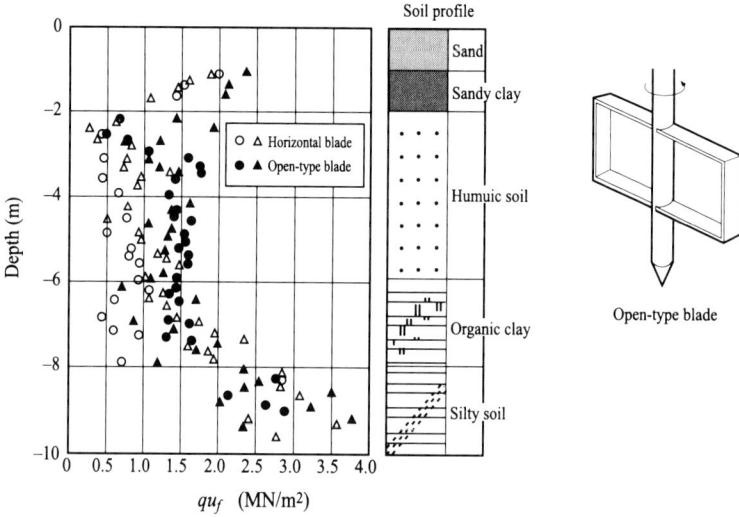

(a) type of mixing blade (Abe et al., 1997)

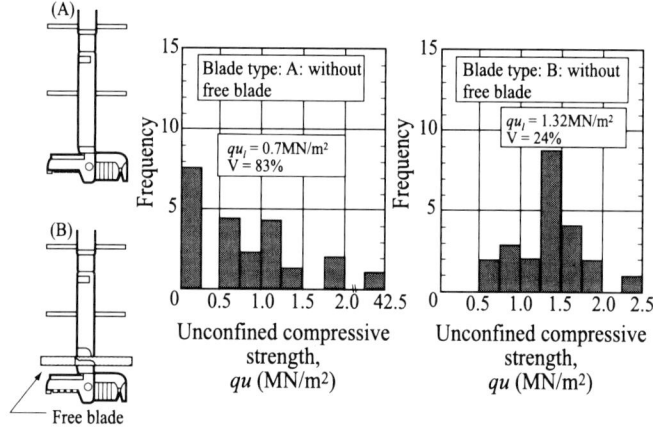

(b) comparison of quality with and without free blade (Enami et al., 1985)

Figure B.2. Influence of type of mixing blade.

Figure B.2 (b) shows the effect of the free blade on the strength, in which two types of mixing blade are compared (Enami et al., 1985). In the mixing blade (B), there is a free blade close to the bottom cutting blade. The free blade is longer than the other mixing blade and is not driven by the rotation, which is expected to stay in the soft soil and to prevent "entrained rotation phenomenon". The figure shows the strength deviation of the in-situ treated soil by the mixing blade (A) is relatively large while those by the mixing blade (B) with the free blade is quite small.

(3) *influence of rotation speed of mixing shaft*. Figure B.3 shows an influence of rotation speed of the mixing shaft, in which two kinds of mixing speed are compared (Nishibayashi et al., 1985). The "blade rotation number " is also different as the rotation speed of the shaft is different. The figure, the strength distribution with the depth, clearly shows that the strength of the treated soil with high rotation speed is higher than those with lower rotation speed. This difference still remained, even in the relatively long curing time.

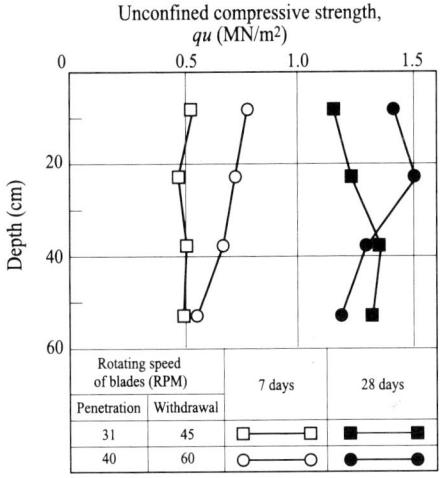

Figure B.3. Influence of rotation speed of mixing blade (Nishibayashi et al., 1985).

(4) *influence of penetration speed of mixing shaft*. Figure B.4 shows the influence of the penetration speed of the mixing shaft on the strength of treated soil (Enami et al., 1986). In the field test, the penetration speed of the mixing shaft is changed from 0.5 m/min to 1.0 m/min while the withdrawal speed of shaft is kept constant at 1.0 m/min. The figure shows that the average unconfined compressive strength decreases very rapidly with increasing the penetration speed of the mixing shaft irrespective of the type of soil. The figure also shows that the coefficient of deviation of the strength becomes large as the penetration speed increases.

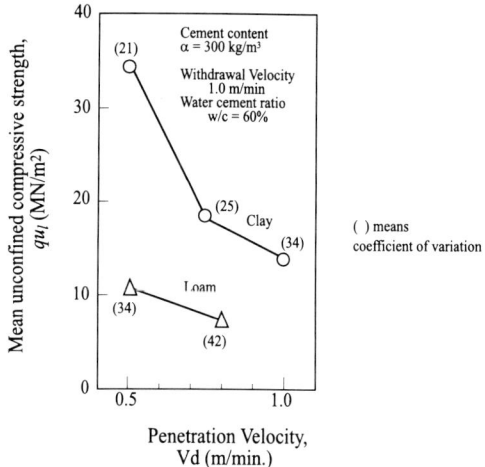

Figure B.4. Influence of penetration speed of mixing shaft (Enami et al., 1986).

(5) *relationship between "blade rotation number" and in-situ strength*. The degree of mixedness of soil and stabilizing agent is one of the most important factors affecting the strength of in-situ treated soil. In Japan, the degree of mixedness is often expressed by terms of the "blade rotation number". The "blade rotation number" refers to the total of rotations of mixing blade passing through a shaft movement of 1 m, and defined by the following equation in the case of the penetration injection method:

$$T = \Sigma M \times \{(Nd/Vd) + (Nu/Vu)\} \quad \cdots\cdots\cdots\cdots\cdots\cdots\cdots \text{Eq. (B.1)}$$

where
    T : blade rotation number (n/m)
    ΣM : total number of mixing blades
    Nd : rotation speed of the blades during penetration (rpm)
    Vd : mixing blade penetration velocity (m/min)
    Nu : rotational speed of the blades during withdrawal (rpm)
    Vu : mixing blade withdrawal velocity (m/min)

Figure B.5 shows the relationship between the "blade rotation number" and strength deviation of in-situ treated soil (Mizuno et al., 1988). The vertical axis of the figure shows the coefficient of variation for in-situ treated soils manufactured by the different blade rotation numbers. This particular field test was conducted to find out the possibility of uniform improvement of a loose sand layer. Among other factors, the influence of blade rotation number is exemplified here. At the blade rotation number of 360, the coefficient of variation ranges between 0.2 and 0.3, which is acceptable strength deviation for most of the practical applications. The figure also indicates the general trend that the deviation decreases with the increase of the "blade rotation number". The similar test data have been accumulated for the improvement of clay soils as well. Based on these evidences, the blade rotation number of 360 is recommended in Japan for wet method of deep mixing.

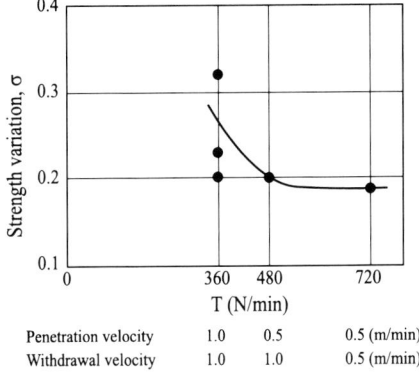

Figure B.5. Relationship between the blade rotation number and strength of in-situ treated soil (Mizuno et al., 1988).

References

Abe, T., Y. Miyoshi, T. Maeda & H. Fukuzumi. 1997. Evaluation of soil improvement by a clay mixing consolidation method using a dual-way mixing system. *Proc. of the 32nd Japan National Conference on Soil Mechanics and Foundation Engineering*: 2353-2354 (in Japanese).

Enami, A., M. Yoshin, N. Hibino, M. Takahashi & K. Akitani. 1985. Basic performance of a mixing method incorporating prevention of rotation of the soil-cement with the mixer- blades. *Proc. of the 20th Japan National Conference on Soil Mechanics and Foundation Engineering* : 1755-1758 (in Japanese).

Enami, A., N. Hibino, M. Takahashi, K. Akitani & M. Yamada. 1986. Soil-cement columns constructed using a compact mixing method machine for housing. *Proc. of the 21st Japan National Conference on Soil Mechanics and Foundation Engineering*: 1987-1990 (in Japanese).

Mizuno, T., Y. Namura & J. Matsumoto. 1988. An experiment on the improvement of sandy soil using the deep mixing method. *Proc. of the 23rd Japan National Conference on Soil Mechanics and Foundation Engineering* : 2301-2304 (in Japanese).

Nishibayashi, K., T. Matsuo, Y. Hosoya, Y. Hirai & M. Shima. 1985. Studies on mixing method improvement of soft ground (Part 4). *Proc. of the 20th Japan National Conference on Soil Mechanics and Foundation Engineering*: 1747-1750 (in Japanese).

APPENDIX C

Recent Research Activity on the Group Column-Type Improved Ground

The group column-type improvement has been extensively applied to improve foundation grounds of embankment or lightweight structure. Recently, Kitazume et al. (2000) performed a series of centrifuge model tests and FEM analyses on the failure pattern and failure envelope of the column type DMM ground, in which a model ground as shown in Figure C.1 was subjected to vertical and horizontal loads. The major conclusions derived in their study are as follows:

1) The column type DMM improved ground fails by one of several failure patterns such as rupture breaking failure of column, domino failure of column and the sliding failure of caisson, depending on the loading condition and the column strength.

2) In the case of the rupture breaking failure pattern, the DMM columns show either shear or bending failure mode depending on the loading condition and their location in the ground.

3) The failure envelopes for the rupture breaking failure pattern are with spindle shapes in the vertical and the horizontal load plane and their sizes are mostly determined by the column strength. The slip circle method incorporating the failure mode, as well as the residual strength of DMM columns, can succeed to evaluate the model tests accurately.

4) The failure envelope for the domino failure pattern is straight line and increases slightly with the increase of caisson weight. The two dimensional elasto-plastic FEM analysis can estimate the failure load reasonably.

5) The stability of the improved ground is described by a combined failure envelope, which is defined by some parts of the failure envelopes giving the minimum capacity.

In this section their test results are introduced briefly.

120 *The Deep Mixing Method – Principle, Design and Construction*

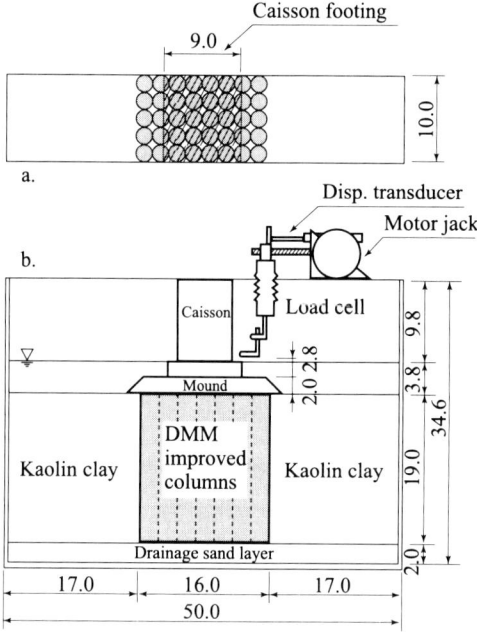

Figure C.1. Schematic view of model ground.

(1) failure pattern of the improved ground. Typical failure patterns of improved ground are shown in Figure C.2. In the vertical loading test (Fig. C.2(a)), the improved ground fails with many rupture breaking failures of column together with several shear failures in some columns just beneath the caisson. The failure patterns observed in the inclined loading tests are also shown in Figures C.2(b) and C.2 (c). The ground with relatively small column strength shows several plain rupture breaking failures with linear character in the columns (Fig. C.2(b)), in which the segments of failed column remain straight. In this case, no clear shear failure mode can be found in the column deformation. The caisson settles deeply into the ground as the columns lose the vertical bearing capacity due to their heavily braking failures. Figure C.2(c) also shows the failure pattern of the improved ground in which the column strength is extremely large. In this case the ground shows the domino failure in which all the columns incline toward the loading direction without any rupture breaking failure.

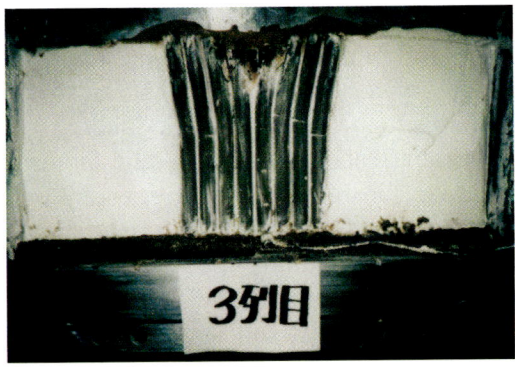

(a) vertical loading (DMMT2)

(b) inclined loading (DMMT12)

(c) inclined loading (DMMT9)

Figure C.2. Failure mode of DMM columns.

(2) *failure envelope*. Figure C.3 shows failure envelope for rupture breaking failure and for the domino failure. The test data are presented with different markers

depending on the failure patterns of ground: full circles for the rupture breaking failure and open triangles for the domino failure. The numbers added next to each marker indicate q_u value of the column in each test. It can be seen that the experiment data classified into the rupture breaking failure form a sort of failure envelope family with spindle shape which has several failure envelopes with the same shape but different size for different column strength.

The test data classified into the domino failure show a comparatively large horizontal load slightly increasing with the increase of vertical load. The envelope for the case of domino failure is determined by elasto-plastic FEM analyses. This envelope represents a straight line (the solid line in Fig. C.3) which passes very close to the test data (open triangles in Fig. C.3). The domino failure envelope increases slightly in horizontal load with the increase of vertical load.

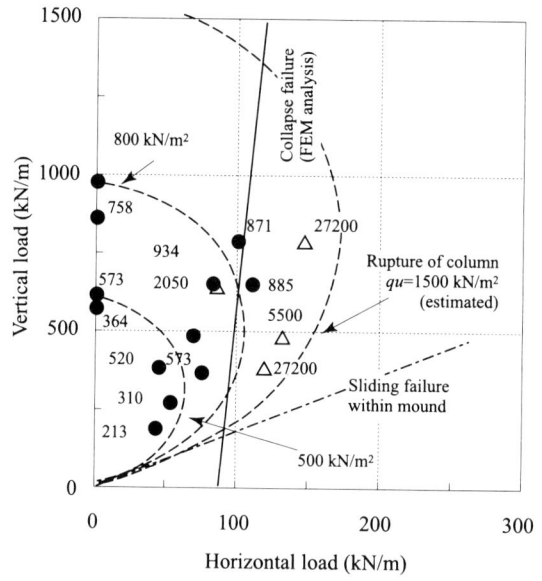

Figure C.3. Failure envelope.

(3) *combined failure envelope*. Three kinds of failure envelopes were found in their tests; the rupture breaking failure, the domino failure and the sliding failure of the caisson. It is reasonable that the improved ground should fail by the failure pattern, which gives the minimum failure load at a particular loading condition. The stability of the improved ground is therefore presented by a combined failure envelope, which is defined by one of the above mentioned failure envelopes giving the minimum capacity. The obtained combined failure envelopes are schematically shown in Figure C.4. When the column strength is relatively small so that the failure envelope for the rupture failure always provides the minimum capacity, the combined failure envelope is the same as the rupture failure envelope and the improved ground always shows the rupture breaking failure (Fig. C.4(a)). When

the column strength becomes relatively large so that the failure envelope for the rupture breaking intersects the domino failure envelope at two vertical loads, V_1 and V_2 (Fig. C.4(b)), the improved ground fails by the rupture breaking failure when the vertical load is larger than V_1 or smaller than V_2, and by the domino failure when it is within the range of V_1 to V_2. As described above, the failure loads for each failure pattern can be evaluated accurately by the proposed calculations.

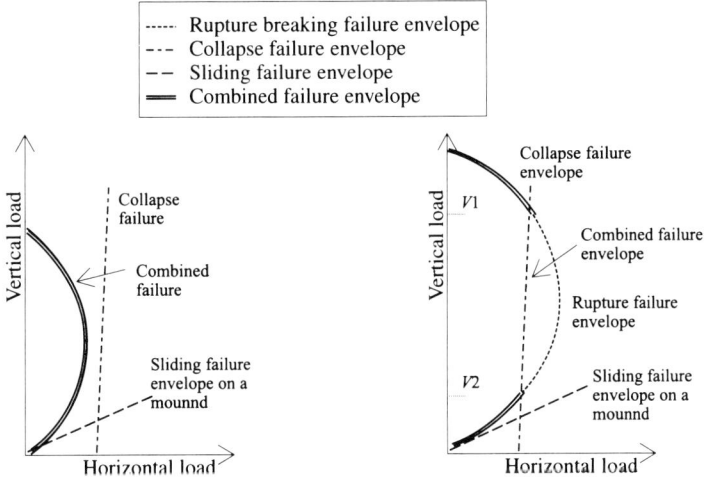

(a) in case of small column strength   (b) in case of large column strength

Figure C.4. Schematic view of combined failure envelope.

Reference

Kitazume M., K. Okano & S. Miyajima. 2000. Centrifuge model tests on failure envelope of column-type DMM-improved ground. *Soils and Foundations*, 40(4) : 43-55.